Stage 2 Design

Electrical Installation Series – Advanced Course

Malcolm Doughton
Charles Duncan
Ted Stocks

Edited by Chris Cox

THOMSON

LEARNING

Australia · Canada · Mexico · Singapore · Spain · United Kingdom · United States

Stage 2 Design

Copyright © CT Projects 2002

The Thomson Learning logo is a registered trademark used herein under licence.

For more information, contact Thomson Learning, Berkshire House, 168–173 High Holborn, London, WC1V 7AA or visit us on the World Wide Web at: http://www.thomsonlearning.co.uk

British Library Cataloguing-in-Publication Data
A catalogue record for this book is available from the British Library

ISBN 1-86152-728-4

First published 2002 by Thomson Learning

Printed in Croatia by Zrinski d.d.

About this book

"Stage 2 Design" is one of a series of books published by Thomson Learning related to Electrical Installation Work. The series may be used to form part of a recognised course, particularly City and Guilds 2360 Electrical Installation Course C, or individual books can be used to update knowledge within particular subject areas. It may also prove valuable for the City and Guilds Courses 2391 and 2400. A complete list of titles in the series is given below.

Electrical Installation Series

Foundation Course

Starting Work
Procedures
Basic Science and Electronics

Supplementary title:
Practical Requirements and Exercises

Intermediate Course

The Importance of Quality
Stage 1 Design
Intermediate Science and Theory

Supplementary title:
Practical Tasks and Revision Exercises

Advanced Course

Advanced Science
Stage 2 Design
Electrical Machines
Lighting Systems
Supplying Installations

Acknowledgements

The author and publishers gratefully acknowledge the following:

AVO International Ltd. (PowerSuite™) for Figures 5.52, 5.53, 5.57, 5.58 and 5.59.
Bussman Division, Cooper (U.K.) Ltd.for Figure 3.4
Menvier (Division of Cooper Lighting & Security Ltd) p.82 Table 4.7
CIP Ltd. 60 New Coventry Road, Birmingham B26 3AY for Figure 8.11.
HMSO for Figure 8.10 and extracts on pp 4 and 107 © Crown copyright is reproduced with the permission of the Controller of Her Majesty's Stationery Office.

Furse (Thomas & Betts) for assistance with material in Chapter 4.
The Institution of Electrical Engineers p.107

Every effort has been made to trace all copyright holders but if any have been inadvertently overlooked, the publishers will be pleased to make the necessary arrangements at the first opportunity.

Study guide

This studybook has been written to enable you to study either in a classroom or in an open or distance learning situation. To ensure that you gain the maximum benefit from the material you will find prompts all the way through that are designed to keep you involved with the subject. If you are studying by yourself the following points may help you.

☞ Work out when, and for how long, you can study each week. Complete the table below and from this produce a programme so that you will know approximately when you should complete each chapter. Your tutor may be able to help you with this. It may be necessary to reassess this timetable from time to time according to your situation.

☞ Try not to take on too much studying at a time. Limit yourself to between 1 hour and 2 hours. When you resume your study go over this same piece of work before you start a new topic.

☞ A project is included in this book. You will find details on p.173. After working through each chapter there will be part of the project to complete.

☞ Try this tasks are included and you may need to ask colleagues at work or your tutor at college questions about practical aspects of the subject. These are all important and will aid your understanding of the subject. You will find answers to the questions at the back of the book but before you look at the answers check that you have read and understood the question and written the answer you intended.

☞ It will be helpful to have available for reference a current copy of BS 7671, IEE Guidance Note 1, IEE Guidance Note 3, IEE Guidance Note 7, Electricity at Work Regulations and the Electricity Supply Regulations 1988. At the time of writing BS 7671 incorporates Amendment No.1, 1994 (AMD8536), Amendment No. 2, 1997 (AMD 9781) and Amendment No. 3 (AMD 10983) 2000.

☞ Your safety is of paramount importance. You are expected to adhere at all times to current regulations, recommendations and guidelines for health and safety.

Study times					
	a.m. from	to	p.m. from	to	Total
Monday					
Tuesday					
Wednesday					
Thursday					
Friday					
Saturday					
Sunday					

Programme	Date to be achieved by
Chapter 1	
Chapter 2	
Chapter 3	
Chapter 4	
Chapter 5	
Chapter 6	
Chapter 7	
Chapter 8	

Contents

6 Estimating 141

7 Planning 149

8 Administration 159

Answers   170

PROJECT
SPECIFICATION   173

Ducan Squash Club

D/001
Ground floor layout, lighting and small power
D/002
1st floor layout, lighting and small power
D/003
Roof level layout, lighting and small power
D/004
Elevation drawings

1

Statutory and Non-Statutory Regulations

Whilst we are engaged in our daily work there are a number of Acts of Parliament and Regulations that we must be aware of. Some of these are statutory which means that they are legally enforceable and failure to comply can result in prosecution. Bearing this in mind it is important for us to make sure that we are aware of these Acts and Regulations and their implications for us.

We shall restrict our considerations in this chapter to some of the Statutory Regulations applicable to our activities. Reference is made throughout this book to non-statutory Regulations, BS 7671 in particular, that we need to consider. You may need to investigate the non-statutory Regulations relevant to other activities such as fire alarms, emergency lighting and hazardous areas.

It is as well to remember that there are Acts and Regulations that relate directly to electricity, its use and associated installations, and others that relate to work and the working environment generally. We shall look at both areas dealing with those related to electricity first.

Figure 1.1

Statutory regulations

These are those backed by Acts of Parliament and are enforced by law.

Health and Safety at Work Etc. Act, 1974

The Management of Health and Safety at Work Regulations 1999

The Construction (Health, Safety and Welfare) Regulations 1996

Construction (Head Protection) Regulations 1989

Lifting Operations and Lifting Regulations 1998

The Construction (Design and Management) Regulations 1994

The Provision and Use of Work Equipment Regulations (PUWER) 1998

The Personal Protective Equipment at Work Regulations 1992

The Manual Handling Operations Regulations 1992

On completion of this chapter you should be able to:

◆ identify the acts and regulations applying to installation and design work
◆ distinguish between statutory and non-statutory regulations
◆ refer to British Standards and codes of practice
◆ demonstrate an understanding of the relevant acts and regulations
◆ identify the responsibilities of all parties under the Health and Safety at Work etc. Act
◆ state the purpose of employment legislation and industrial agreements relevant to your position

Factories Act 1961 and the Regulations made under this act

Construction (General Provisions) Regulations 1966

Offices, Shops and Railway Premises Act 1973

Occupiers Liability Act 1984

Control of Substances Hazardous to Health Regulations 1988 (COSHH)

Noise at Work Regulations 1989

Workplace (Health, Safety and Welfare) Regulations 1992

Electricity Supply Regulations 1988

Electricity at Work Regulations 1989

Building Standards (Scotland) Regulations 1981

Agriculture (Stationary Machines) Regulations 1959

Cinematograph Regulations made under Cinematograph Acts of 1909 and 1952

*The Highly Flammable Liquids and Liquefied Petroleum Gases Regulations 1972

*The Petroleum (Consolidation) Act 1928

*These last two come under the Home Office.

Remember
All legislation is subject to change and when making reference to any Act or Regulation it is important to ensure that ALL the amendments are included. Information on the current status of each can be obtained from HMSO or the relevant authority.

Some of these Acts and Regulations have particular reference to certain areas of the industry, for example the Electricity Supply Regulations 1988. We cannot discount these as they affect the nature and level of electrical supply we may expect from the companies supplying electricity.

Let us consider the Electricity Supply Regulations 1988 with particular reference to some of those which have a bearing upon our electrical installation.

The Electricity Supply Regulations

It is worthwhile obtaining a copy of the Electricity Supply Regulations 1988, including all amendments, and reading through the exact requirements placed upon the supply companies and the consumer. This will also affect us as the designer/installer of the installation. We are the experts engaged by the consumer to ensure that the installation complies with all the relevant Acts and Regulations.

The Regulations are extensive and obtaining a copy for reference is the best way to fully appreciate the requirements. We shall consider those which have the most significance to the consumer and installer.

Regulation 7
Paragraphs 7, 8 and 9 detail the requirements for earthing of metalwork, circuit protective conductors and bonding within the consumer's premises when connected to a TN-C-S system (PME). Paragraph 9 sets out the minimum c.s.a. for bonding conductors in this situation. We should remember that this refers to main bonding conductors and not to supplementary equipotential bonding which is covered by BS 7671, Requirements for Electrical Installations.

Regulation 27
Paragraph 1 refers to the fact that the supply company is not obliged to commence, or to continue to provide, a supply unless they are satisfied that the consumer's installation is constructed, installed, protected and used, so far as is practicable, to prevent danger. The supplier should also be satisfied that the consumer's installation will not cause undue interference to the supplier's system or to other consumers. Paragraph 2 tells us that any consumer's installation which complies with the requirements of BS 7671 shall be deemed to comply with the requirements of this Regulation (Regulation 27 of the 1988 Act) as to safety.

Regulations 28 and 29

These refer to the procedures for disconnection of supply where the supply company is satisfied that the consumer's installation, or a part of it, does not comply with Regulation 27. These Regulations also refer to the requirements for notification and the resolution of differences between the supply company and the consumer.

An installation which only complies with the requirements for safety in the Electricity Supply Regulations will not necessarily comply with BS 7671, whilst installations to BS 7671 will comply with the requirements for safety to the Electricity Supply Regulations.

Regulation 30

This covers the declaration by the supply company of;

- the number and rotation of phases
- the frequency and
- the voltage

at which it intends to supply and the extent of the permitted variations of the declared values. These declarations should be made prior to the connection of the supply.

Regulation 31

This details the information that the supply company must provide, on request, to the consumer as a written statement. These are

- the maximum prospective short circuit current at the supply terminals
- the maximum earth fault loop impedance of the earth fault path external to the consumer's installation
- the type and rating of the supplier's fusible cut out, or switching device, nearest to the supply terminals

These values that apply, or will apply, to that installation should be provided to any person who can show reasonable cause for requiring this information.

We have only touched on the requirements of the Electricity Supply Regulations 1988 but it is important that you are fully aware of all the requirements. To this end some further investigation is required on your part. As we progress through the project contained in this book, we may need to refer to the requirements of the Regulations.

The Electricity at Work Regulations 1989

Let us now consider the requirements of the Electricity at Work Regulations 1989 which came into force 1st April 1990 and are issued under the Health and Safety at Work Etc. Act 1974.

We must first consider the scope of these Regulations, and a brief summary is that they

- do not limit their application by voltage level and apply equally to systems operating at any voltage, for example from a battery powered hand lamp to a 400 kV line
- impose duties and responsibilities on employers, employees and the self employed
- apply anywhere within the UK where work takes place on or near electrical systems.

To make it easier to consider the requirements we can split the Electricity at Work Regulations 1989 into groups which apply to certain areas.

- Definitions, interpretation and duties are covered by Regulations 1 to 3.
- Safe systems of work, competence and training are dealt with in Regulations 4, 11, 12, 13, 14, 15, and 16.
- Installations, equipment and their environment are covered by Regulations 5, 6, 7, 8, 9 and 10.
- Mines and quarries only are referred to in Regulations 17 to 28 inclusive.
- Regulation 29 is the defence Regulation
- Exemptions, modifications and extensions are covered in Regulations 30, 31, 32 and 33.

The aim of the Electricity at Work Regulations is to establish a safe working system relating to the life of an electrical system. To achieve this they impose legal duties and requirements upon

- Design and planning
- Installation and construction
- Commissioning
- Use
- Routine and post fault inspection and repair
- Disassembly and disposal at the end of useful life.

The Electricity at Work Regulations also make it an absolute duty to prevent danger; and danger is defined as "a risk of injury". The qualifying statement "as far as is reasonably practicable" is used in many Regulations. This means that some level of judgement may be applied, based on the risk of danger against factors such as time or cost. It must be appreciated that the higher the risk of danger the less importance can be attached to factors such as cost. If no such qualification is offered in any regulation then the regulation is an absolute requirement and must be complied with regardless of any other factors.

The Electricity at Work Regulations place legal duties on every person involved in electrical work by virtue of their relating to matters under that person's control. We have the legal requirement placed upon us to take adequate safety precautions to prevent injury from the dangers of electricity, whether working on live or dead conductors.

It follows therefore that any electrical accident is likely to involve a breach of the regulations. As the regulations are statutory such a breach leaves the company and individuals concerned liable to prosecution. In which case the burden of proof that there was no breach of the regulations rests with those accused, guilty unless proven innocent.

It is as well to acquaint ourselves with the defence regulation No. 29.

Regulation 29

"In any proceedings for an offence consisting of a contravention of regulations 4(4), 5, 8, 9, 10, 11, 12, 13, 14, 15, 16 or 25 it shall be a defence for any person to prove that he took all reasonable steps and exercised all due diligence to avoid the commission of that offence."

Remember
If no "as far as is reasonably practicable" qualification is offered in a regulation it is

AN ABSOLUTE REQUIREMENT

and must be complied with regardless of any other factors.

It is a requirement of the Electricity at Work Regulations that anyone involved in any electrical activity must have the necessary technical knowledge or experience to avoid danger. If this is not the case then the individual must be under the appropriate supervision. This is generally intended to allow people to be trained in new activities rather than to allow anyone to carry out activities in the presence of supervision.

For example a group of unskilled operatives working under one skilled supervisor is not acceptable. On the other hand the training of an operative, such as an apprentice or a skilled individual in a new task, is typically an acceptable situation.

Let's take a look at some of the more relevant regulations, these being those that have implications for people that use electrical systems.

Regulation 4
This states the general requirements for the construction and working of a system. It is as well to note that "construction" incorporates design, commissioning, use and maintenance and the "work" includes switching and testing as well as electrical work.

Regulation 13
This establishes the absolute requirements for working on electrical equipment that has been made dead. If we are to comply with this regulation we will have to establish a safe system of work which will ensure we achieve the following:

- isolation from all points of supply
- secure (lock off or remove fuses) each point of isolation
- ensure that equipment is proved dead at the point of work
- identify safe areas of work
- provide barriers from adjacent live equipment
- provide and operate a permit to work system

Regulation 14
This states the conditions that must be met in order to work on live equipment and prevent injury and this Regulation is an absolute requirement. It is also concerned with the justification for working live.

Regulation 16
This relates to the need for those who work on electrical systems to have adequate knowledge and/or experience for the activities or to be appropriately supervised to prevent danger or injury. So to fulfil this requirement we must have the following level of knowledge or experience

- adequate knowledge and understanding of electricity
- appropriate experience of electrical work
- understanding of the system/equipment to be worked on and practical experience of that system/equipment
- understanding of the hazards that may arise during the course of the work and the precautions that need to be taken
- the ability at all times to recognise if it is safe for work to continue

Regulation 29

This provides the defence that all due diligence was exercised and all reasonable steps taken in relation to the Regulations which are an absolute requirement.

Try this
List all the Electricity at Work Regulations that are an absolute requirement.

Introduction to the project

In the appendix at the end of this book there is a project for you to complete. This contains tasks and activities which are important elements in the design of an electrical installation. It is therefore important that you complete the project as a self assessment exercise. The requirement to complete a section of the project occurs at the end of most chapters within this book.

In order to assist you in this task we shall consider the part of the project which should be completed at the end of each chapter. In this way we shall complete the project in stages with each stage being related to the chapter just completed.

Part 7 of the project will require you to consider the fire alarm requirements for the building. This section is intended for you to familiarise yourself with the relevant British Standards. It must also be appreciated that this is one area that may well feature in your City and Guilds project.

The inclusion of this section is to provide you with a self researched task, as this sort of problem may well occur in your everyday work with areas new to you. These will then need to be investigated in order to ensure that the installation is correctly carried out and complies with the relevant standards and codes of practice.

The project comprises a specification, a set of drawings and a list of tasks. There are two drawings of the ground and first floors, a plantroom layout and an elevation which are not to scale, and a legend for the symbols.

If you are using this book as part of your studies for the City and Guilds "C" Course you will find that the project covers topics which are vital elements of the course. It is helpful to complete the project as part of your preparation for the City and Guilds project and the written examination. Whilst the project contained within this book involves some tasks which are similar to those which may be set in the City and Guilds project, it is not intended to be used as an alternative to that required by the City and Guilds.

Good Luck!

PROJECT

Now we have completed Chapter 1 of the module we should consider parts 1a and 1b of the project. These deal with the statutory and non-statutory requirements for our installation activities. In order to complete this section you will need to refer to other documents such as the "Electricity at Work Regulations" so make sure you have these available before you begin.

2

Heating Systems

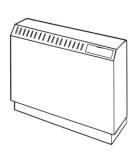

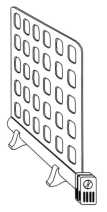

Figure 2.1

On completion of this chapter you should be able to:

◆ describe the different types of electric heating systems
◆ describe the methods of wiring, control and protection used for thermal storage systems
◆ state the advantages and disadvantages of different systems
◆ list the methods used for electric water heating
◆ produce circuit diagrams for the external control wiring for heating systems using other fuels
◆ calculate the electrical power requirements for various heating schemes
◆ select and site the most suitable form of electric heating for given situations

We shall begin this chapter by considering the different forms of electric space heating and water heating. Into this category come the various types of heating from the traditional electric fire to the most sophisticated storage system. In order to fully appreciate the different types we must first be aware of the different methods of heat transfer and look at the types of heating appliance and systems utilising these various methods.

In general terms heat is transferred from one medium to another by the following methods:
• Radiation
• Convection
• Conduction

Radiation

This is where heat is radiated out from a hot body. The traditional coal fire operates on this principle as does the good old radiant electric fire. If a material such as coal is burnt then heat is produced. The most common way for this heat to be transferred from the coal to the room is by radiating the heat outwards in much the same way as light is radiated from a light source.

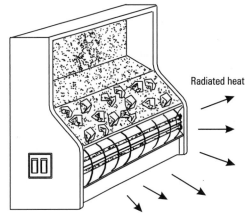

Radiated heat

Figure 2.2   Radiant fire

If you have ever stood in front of a coal fire you will also be aware that radiated heat follows very closely the characteristics of light radiation. For example the further away from the fire you stand the less heat reaches your body and this almost follows the inverse square law as applied to light.

A further similarity is that the side closest to the fire is nice and warm, the side furthest away stays cold, so we know that radiated heat does not bend round corners and can be shadowed just as light can. We can also demonstrate that the air through which the radiated heat passes is hardly affected at all and the air temperature is subjected to a very small increase. In much the same way as light falling on a surface causes the surface to be lit but does not light the air through which it passes.

Convection

This is where heat is transferred from the heat source by the movement of another medium such as air or oil. If we consider the "flow" of oil within a transformer tank, we can get some idea of the principle involved.

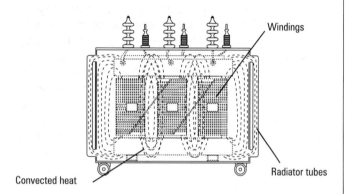

Figure 2.3   *Heat convected through the insulating oil of a power transformer*

In Figure 2.3 we can see that the oil is heated by the transformer coils dissipating the heat produced by current flow. Hot oil rises through the tank towards the top and as it does so is replaced by cool oil falling into its place. As the oil rises through the tank some heat is dissipated into the surrounding oil and transferred to the outside air through the radiator tubes. As the oil cools it begins to fall through the tank and down towards the bottom. On reaching the bottom the oil is drawn up through the "hot" area, it begins to rise again and so the cycle continues.

Whilst this is only an approximation of the principles involved it serves fairly well as an analogy to the behaviour of convected air within a room. A heat source placed within a room will produce a similar pattern in the flow of air within the space, due to its heating effect.

In Figure 2.4 we can see the result of this air cycle in diagrammatic form and as the function produced is, to a degree, controlled by natural forces, the heating element will produce a gradual rise in the air temperature throughout the room.

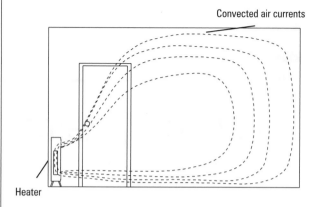

Figure 2.4   *A convector heater used to heat a room.*

Conduction

Conduction is best described as the transfer of heat through a solid body. If, for example, we place a poker in a fire and leave it in for some time the end of the poker will become red hot. At the same time the handle will become warm and if left for long enough may become too hot to hold. The transfer of heat from the fire to the poker is, by and large, the product of radiation, whilst the transfer of heat from the hot end of the poker to the handle is due to conduction.

For our purposes the most significant examples of conduction are found in four main areas:
- from cable core to cable surface through the insulation
- from an immersed element to the surrounding medium, for example kettle element or immersion heater
- storage heaters and underfloor heating systems
- heat loss from buildings through the building structure

Summary
Although we have considered the 3 main methods of heat transfer as separate items they are, however, almost impossible to isolate to this degree. If we use the oil filled electric radiator as an example, we can see why this is so. The element itself is heated by the passing of current through a special conductor and the heat produced is conducted to the outer wall of the element. From this point the heat is conducted through the oil to warm both the oil and to some extent the exterior metal at the base of the radiator. The oil then transfers the heat through the radiator by convection until all the oil is hot. The heat is transferred by conduction to the metal of the radiator. At this point the majority of the heat is transferred to the air by convection although if you are close to the radiator you will feel heat being radiated, hence the name. As a final aside, some heat will also be conducted, through whatever supports are used, into the building fabric.

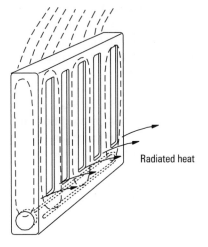

Convected heat

Radiated heat

Figure 2.5   Electrically heated oil filled radiator

Space heating methods

It would be a good idea for us to look at the types of space heating systems currently employed and to consider the principle and advantages of each. It is important to remember that when we refer to a space heating scheme we mean a system which provides the total heating for an area. In the case of the single electric fire in the lounge, which is used when the temperature drops a little and the gas boiler is not switched on, does not constitute a space heating system, it is merely a top up. A set of night storage radiators installed throughout a domestic dwelling would constitute a full system. The same electrical fire installed in the lounge would still be in addition to the main heating system and so would not form part of that system.

Radiant

Direct heating radiant systems tend to be inefficient for space heating and are generally restricted to little used areas with infrequent occupation. In locations such as station waiting rooms it is quite common to use infrared heaters controlled by touch control timed switches. This method of control ensures that the heaters are only switched on when the room is occupied and the timed switch-off facility makes sure the heaters are not left on unnecessarily. This type of installation is typical for many public access applications occupied for short durations.

Remember
Heat always passes from a hotter body, substance or medium to a colder one and NOT from colder to hotter.

Try this
For each method of heat transfer select two appliances which employ that principle as the main method of heat transference.

Radiation
Appliances:

Convection
Appliances:

Conduction
Appliances:

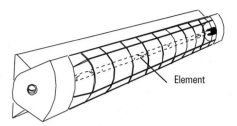

Figure 2.6   Radiant infrared type heater

Direct radiant heaters are often employed in specialist locations and halogen-based units are commonplace. For example they are used in catering to keep food hot in sales display cabinets. Whilst these halogen-based units are more efficient than the old radiant heater they are still relatively inefficient for area heating. However, they have a number of advantages for particular applications, some of these being that

- they are instantly available without warm up
- they keep objects warm without raising the surrounding air temperature
- they can be accurately focused to ensure that heat is delivered where required in the most efficient and effective manner
- output can be varied by the use of simple control equipment for example a dimmer switch

They also have the advantage of providing a heat source that is effectively shrouded by glass thus reducing the risk of contact with an exposed element. One typical application would be a bird rearing station where the heat can be focused directly onto eggs or young without risk to the surrounding straw medium and the risk to the livestock of an exposed element.

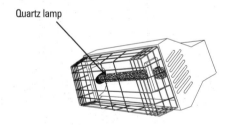

Figure 2.7   Linear quartz heater

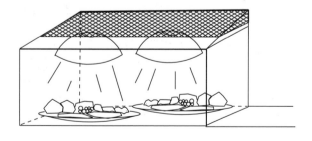

Figure 2.8   Halogen heated food display

Fan heaters

It would be as well at this point to consider the use of fan heaters as these do have a place in area heating. It has become a recognised practice to use wall-mounted Class 2 fan heaters, controlled by a cord switch, in bathrooms. This provides a method of almost instant heat without the risk of exposed elements or metallic cases.

Fan heaters are also used as a means of providing additional heat in specific areas when conditions dictate. They will frequently be found in use, in the form of portable appliances, around offices and homes during the winter months.

Larger fan heaters are often used to provide a "curtain" of hot air at doorways. As the door is opened and closed warm air is lost to the outside and is replaced by cool air entering the room or building. The air curtain provides some of the warm air lost to the outside and offers a barrier to prevent the ingress of cold air.

This system of heat loss prevention is often employed in shops as there is continuous traffic in and out. The cost of running a relatively small fan heater is outweighed by the saving on the full heating system. Improved user comfort is an added bonus.

The principle of operation of a fan heater is that a small fan, usually of cylindrical construction, forces air over an element. The element can be totally contained within the case and therefore inaccessible. This is in essence a forced convection heater, but we are considering it at this time because one common usage for this type is in areas where it replaces, or supplements, the conventional radiant heater. A typical fan heater is shown in Figure 2.9.

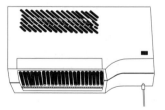

Figure 2.9   Wall mounted fan heater

Convection

Convector heaters fall into a number of categories so we'll look at some typical types and applications.

Low wattage units: these are generally used to maintain a relatively low ambient temperature in areas such as greenhouses, stores, sheds and airing cupboards. These units are usually self regulating containing inbuilt thermostats and are left on continuously under their own control.

Larger oil filled radiators are often used to heat bigger areas, again under the control of their own integral thermostat. They are often employed in bathrooms as they provide a reasonably effective, if not altogether economic, heat source with no exposed elements. Provided they are suitable for the environment and correctly installed and supplied via an outlet plate, they can comply with the requirements for electrical installations within bathrooms.

These units could be used to provide a full space heating system. They are, however, not very efficient and whilst they can raise larger room temperatures their make up rate due to losses, for example opening a door, are so slow that such a system would be both inefficient and ineffective.

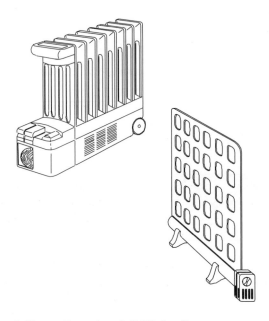

Figure 2.10   Examples of oil filled radiators

Storage convector heaters are widely used for domestic installations to provide a full space heating system. This heater in its simplest form is shown in Figure 2.11 and consists of a number of fire-bricks with an electric heating element run through them. When the supply is connected to the element the bricks are heated up and, once fully heated, these retain the heat for some time. By surrounding them with insulating material, such as rockwool, the heat can be retained for a longer period. Airways are constructed through the bricks and then sealed with a damper. This gives the facility to control the flow of air though the blocks and therefore the rate at which heat is dissipated. The heat is carried into the room by natural convection. This type of installation requires a heater of this type in each room.

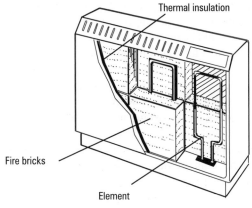

Thermal insulation

Fire bricks

Element

Figure 2.11   Storage radiator

The main advantage of this type of heater is that, due to its ability to store heat for long periods, it can be operated on "off peak" electricity. This allows the electricity supplier to level out the demand over the 24 hour period and, because of the saving in plant utilisation, the energy is sold on to the consumer at a reduced rate. In this way the electricity supplier and the consumer both benefit from this type of system which is commonly known as the "economy seven" tariff. This is so named because the electricity company undertakes to supply consumers with reduced cost electricity for at least 7 hours in any 24. With modern metering and control methods this need not be a consecutive time period but can be broken down into smaller, more frequent periods depending on demand.

Fan assisted storage systems are a variation on the conventional storage radiator insomuch as a single large storage unit may be installed, usually in a central location. The storage blocks may be made from cast iron instead of firebrick and airways are constructed within the blocks in the same way. The unit has air plenums with ductwork running from one plenum to supply grill outlets in each room rather like a mechanical ventilation system. The other plenum provides a housing for intake air grille(s) and filters connected to a supply air fan as shown in Figure 2.12.

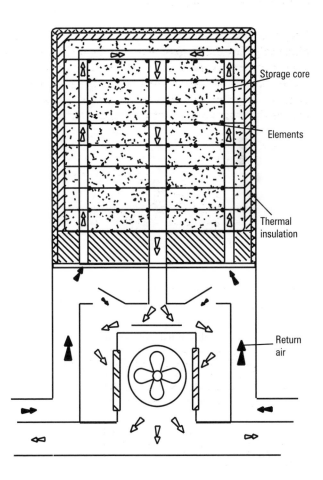

Storage core

Elements

Thermal insulation

Return air

Figure 2.12   Simple fan assisted heat storage system

The thermal blocks are heated in the same way as those for the convector radiator. In order to discharge the heat into the rooms the fan is run drawing air through the hot blocks and into the ductwork. The system must be balanced to ensure that the warm air is distributed to each room in the required quantities.

Once expelled into the room the air circulates by natural convection and is drawn back to the fan through return grilles. This system also makes use of the cheaper energy rate offered by the electricity supplier. This type of heating system may be used in industrial/commercial situations quite successfully. The physical requirements of the unit and airways mean, for domestic use, it has to be built in at the time the dwelling is constructed. It is not normally possible to install such a unit in a previously completed dwelling.

Underfloor heating was used by the Romans in their villas, nowadays we can install underfloor electric heating in commercial and domestic properties. A special heating cable is installed in the final floor screed as shown in Figure 2.13.

Once the installation is complete and connected up the whole floor becomes, in effect, a large storage heater. As heat rises the rooms are warmed by natural convection. This is another system that uses the off peak facility offered by the electricity supplier.

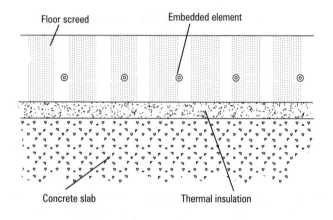

Figure 2.13   Underfloor heating

It is important to identify areas of floor where fixed equipment is to be located, or where equipment will need to be fixed to the floor, as heating cables must not be installed where these situations occur.

One of the main advantages of this system is that the heating appliance does not occupy any space within the finished room.

We may use this type of system to provide soil warming for horticultural applications or used to keep ramps and roadways free from ice.

Ceiling heating

This arrangement has proved to be less popular than the underfloor heating but is worth consideration. Each room is fitted with a heating element within the fabric of the ceiling structure and these elements radiate heat down into the room. It is normal practice to have each room controlled by an individual thermostat. This system, as with the floor warming system, occupies no space within the finished room.

Figure 2.14 shows a typical layout.

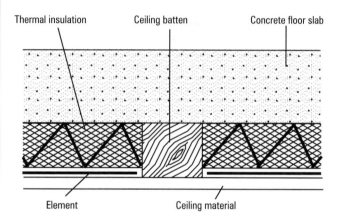

Figure 2.14   Ceiling heating

Both ceiling and floor heating systems employ relatively low output heating elements, typically between 16 and 18 watts per metre. This means that the actual surface does not become unbearably hot to the touch. The output can be varied by adjustment of the spacing between the element loops in accordance with the manufacturers' recommendations. Ceiling heating elements are normally supplied in standard widths to suit joist centres, but these can be altered at the factory or fabricated on site to accommodate non-standard layouts and changes.

Combination of types

We have considered different heat sources and the relevant types of heaters and by now you could probably think of applications where more than one type of heater would be advantageous. It would be nice to have a radiant heater in the lounge with the facility to blow warm air into the room to raise the ambient temperature on autumn evenings, for example.

We could find several instances where a storage radiator would be appropriate but we would also like to supplement this with an additional panel or convector heater for on peak use. A facility to speed up the rate at which the storage heater gives heat into the room, by supplementing with a fan heater, could also be an advantage.

Many of the manufacturers are aware of these requirements and produce units to provide features such as those mentioned. Some examples are shown in Figure 2.15.

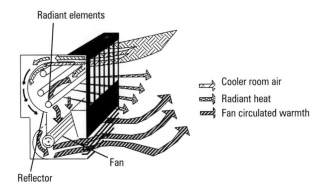

Cooler room air
Radiant heat
Fan circulated warmth

Radiant heater with fan assistance

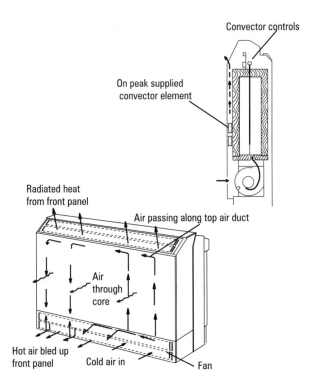

Figure 2.15

Storage heater with on peak convector

Heat recovery now plays an important role within commercial installations. Within a typical office block, there is a considerable amount of heat generated by people, equipment and lighting. Many schemes are now being installed to recover the heat produced by light fittings and recycle this back into the building. A typical arrangement for one such system is shown in Figure 2.16.

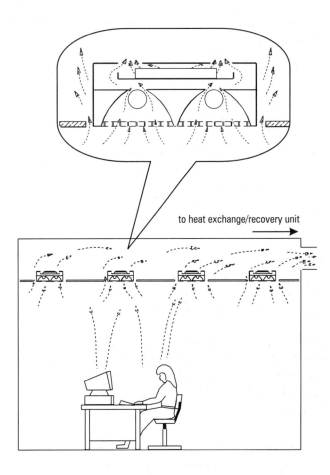

Figure 2.16   Heat recovery

Heat recovery does require the control of the building environment, air movement etc. Therefore it is uncommon for this type of system to be used in domestic environments but schemes have been developed for domestic heat recovery.

Many new buildings for commerce and business premises are using heat recovery and it is becoming commonplace in major refurbishment projects.

A heat recovery unit is based upon extracting heat from expelled stale air or water and transferring it to fresh air intakes. The simplified layout and principle is shown in Figure 2.17. Figure 2.18 shows how the unit could be installed within a building to recover heat from extracted air. Should we wish to recover heat from light fittings then this can be achieved either by purpose made ducting or by making the ceiling into an air plenum from which we recover heat by a similar process to that described above.

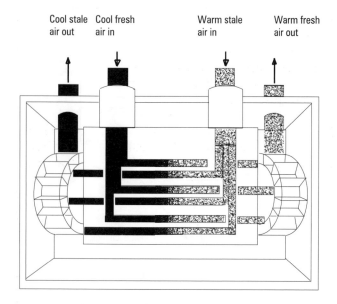

Cool stale air out | Cool fresh air in | Warm stale air in | Warm fresh air out

Figure 2.17   Heat recovery unit

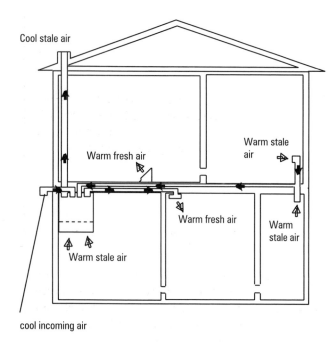

Cool stale air

Warm fresh air

Warm stale air

Warm fresh air

Warm stale air

Warm fresh air

Warm stale air

Warm stale air

cool incoming air

Figure 2.18   Typical heat recovery distribution

Tariffs

Several of the above systems make use of the off peak tariff offered by the electricity suppliers. Perhaps a recap of just what is involved is required in order to appreciate why these tariffs are offered and just what their effect is on both consumer and supplier.

The main motivation behind the offer of an off peak tariff is to even out the peaks and troughs in the demand for electricity. Figure 2.19 shows a typical demand graph of a 24 hour period, suitably simplified to enable us to examine the principles involved.

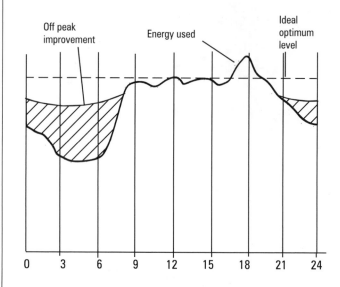

Off peak improvement   Energy used   Ideal optimum level

0   3   6   9   12   15   18   21   24

Figure 2.19   The electricity demand over a typical 24 hour period

We can see that during the period from around midnight to about 7.00 in the morning there is a pronounced trough in demand. The nature of generating equipment is such that it is not practical or possible to start and stop generators in response to demand with the possible exception of diesel sets and hydro-electric schemes. The cost of running a generator on small load is pretty much the same as running it at high load. Therefore, running at low load represents a considerable fall in efficiency for the generating company.

By offering consumers electrical energy at a much reduced cost during the periods of low demand the electricity suppliers have managed to increase the use of electrical energy in these periods and improve operational efficiency. The main use for off peak electricity is in heating appliances of the thermal storage type.

The type of off peak tariff offered depends on the type of consumer, and the electricity suppliers are always pleased to advise consumers of the most suitable and cost effective tariff for their individual needs. Table 2.1 shows some typical tariffs offered but for the most recent information it is advisable to contact the electricity suppliers. As the choice of supplier is quite large it is worth contacting several suppliers to establish who offers the most beneficial package for a particular consumer.

Table 2.1

Typical domestic tariffs

Economy 7 domestic tariff

Quarterly Charge	£13.23
Unit charge – Day rate	6.237p
Unit charge – Night rate	2.573p

Standard tariff

Quarterly Charge	£8.86
Unit charge – Day rate	6.080p

Business tariffs

Economy 7 business tariff

Standing Charge	£17
Unit charge – Primary units	6.34p
Unit charge – Secondary units	6.08p
Unit charge – Night rate	2.45p

Standard business tariff

Standing Charge	£14
Unit charge – Primary units	6.05p
Unit charge – Secondary units	5.80p

Evening, weekend and night business tariff

Standing charge	£17
Unit charge – Primary units	7.65p
Unit charge – Secondary units	6.95p
Weekend and Evening units	3.13p

Three-Phase supplies

Additional Quarterly Charge for Three-Phase supplies on Business Tariffs	£5.37

Note: Primary units usually apply to the first 3000 daytime units used each quarter.

Try this

For the following consumers obtain the most recent tariff information from your current electricity supplier and suggest the most suitable tariff in each case.

1. A domestic consumer with a three bedroom home which is to be heated by nightstore radiators. The total storage heater load is 12 kW and an option for a 3 kW immersion heater is available.

2. A commercial consumer who leases out a small office block which is heated by fan assisted electric storage units rated at 18 kW each and has a total of six such units. The offices are not occupied between the hours of 18.00 and 08.00.

3. A domestic consumer with a two bedroom home which is to be heated by a gas fired boiler. The 3 kW immersion heater is available for use during the summer months. The family have no children.

4. A commercial consumer who has a maximum demand of 250 kVA and whose premises are heated by gas fired boilers. The works premises operate on a three shift system and so plant and equipment is operational 24 hours each day.

Water heating

Electric water heaters are used in a wide variety of domestic and industrial applications. With many different heaters available correct selection is important but, generally speaking, these fall into two main categories; storage and instantaneous types. We shall consider some of the types of heater available and look at their most suitable applications, beginning with those found in domestic installations.

Storage types

This is one of the most common forms of water heating to be found in a domestic installation and is usually in the form of a large hot water tank fitted with an immersion heater normally supplying all the hot water taps in the house. A simplified version is shown in Figure 2.20

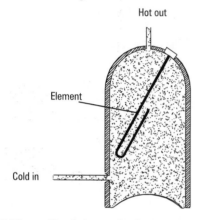

Figure 2.20   *Simple immersion heater*

Storage water systems such as these have the advantage of making a large quantity of hot water available at any one time and usually with a fast delivery rate. They can also take advantage of the off peak electricity tariff. Their main disadvantages are that once the stored water is used up they have a long re-heat time and they are subject to heat losses when water is not being used.

To help overcome these problems some manufacturers have a version that has two immersion heaters as shown in Figure 2.21.

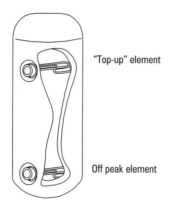

Figure 2.21   *Dual tariff immersion heater*

These are usually connected so that the lower element is supplied during off peak times and the whole tank is heated. The top element is only used during high rate times when the tank's hot water has been used. This element tops up the tank until the off peak element comes back on via the time switch.

Smaller storage heaters are sometimes found located either above or below a sink supplying a single tap, as shown in Figures 2.22 and 2.23.

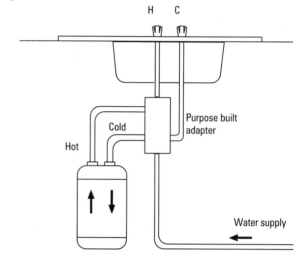

Figure 2.22   *Undersink water heater*

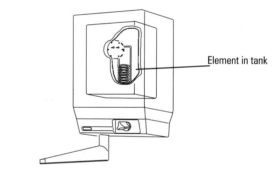

Figure 2.23   *Oversink water heater*

Instant types

The electric shower is a common feature in many homes and operates on the principle of a restricted flow of water through a tank containing a high power element (Figure 2.24).

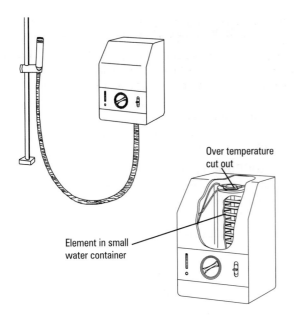

Figure 2.24   *Shower unit*

A smaller version is used to provide hot water to sink taps (Figure 2.25).

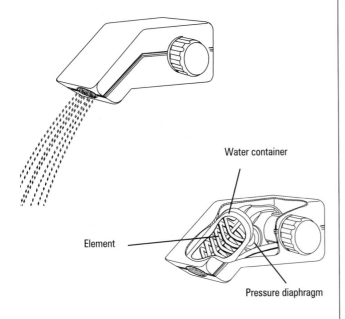

Figure 2.25   *Hand washing unit*

This type of water heater has the advantages of being economical, as water is only heated when it is required, and is able to supply unlimited amounts of hot water, within reason.

Their principle disadvantages are that the flow rate is considerably less than that from a storage system, they can usually only supply one tap and in the event of a power failure no hot water is available. Whilst this is also true of the storage types, with the instantaneous types the lack of hot water is also immediate and of course they cannot take advantage of the off peak tariff as the storage type does.

Electric water heaters which can be used to supply a "wet" central heating system are also available. They have virtually no storage capacity and rely on the pumped circulation of water through the secondary system to dissipate the heat generated. Unless they are supplied with off peak electricity at reduced rates, they are expensive to run in comparison with other fuels. Applications that are sometimes encountered are heating small swimming pools and protection of areas from frost damage. A typical system is shown in Figure 2.26 with the electrical control schematic in Figure 2.27.

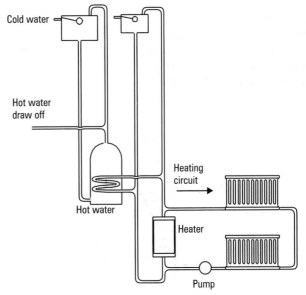

Figure 2.26   *Electric "wet" heating system*

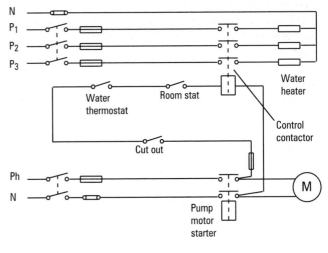

Figure 2.27   *Electric "wet" system diagram*

Industrial water heaters

The types of water heater available for industrial use are essentially the same as those found in the domestic installation. The main variations are in the capacity and the method by which they actually heat the water, both of which are dependant on their application.

Instantaneous types

The small domestic type, using the immersion element, is used quite extensively for hand washing facilities. For many applications however this method provides a flow rate too low to be of any practical use.

To overcome this electrode type boilers are used. These can be quite small and often feature in hot drinks machines. They have the advantage of offering a quantity of hot water readily available for use and a very fast heat up time.

The principle of an electrode boiler is that electrodes are placed in the water and a current is passed between them through the water. The passage of current through the water produces heat and it is this heat production that causes the water temperature to rise. A simple arrangement is shown in Figure 2.28.

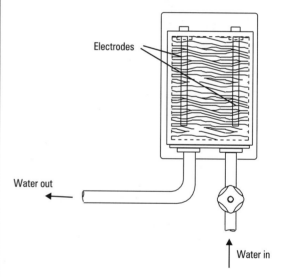

Figure 2.28   Small "instantaneous" electrode heater

This sounds like a fairly hazardous business and this type of unit is subject to particular special requirements as can be seen later. BS 7671 makes particular reference to the installation of electrode boilers.

Much larger electrode boilers are employed where large quantities of hot water are required and whilst these may be three-phase and operate above low voltage their principle of operation does not alter.

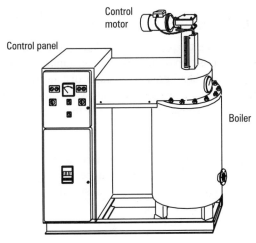

Figure 2.29   Electrode boiler

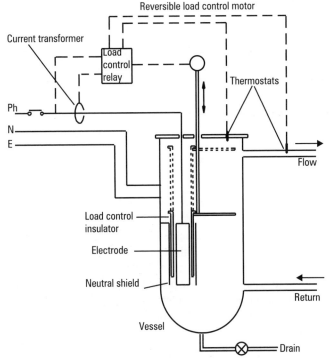

Figure 2.30   Electrode boiler diagram
                (Only one phase and electrode is shown for
                clarity)

Electrode boilers have been used extensively throughout the world since their introduction in the 1930's. They are available in sizes up to 3100 kW and are used instead of conventional boilers. The advantages that they have over fossil fuels are that they require no chimney, no fuel store, no special fire precautions, require little maintenance and can be operated by unskilled persons.

The current source for the type of boiler shown in Figures 2.29 and 2.30 is via electrodes in multiples of three, connected to a three-phase supply. Each electrode is made from a close grained cast iron alloy and is surrounded by a cylindrical neutral shield. Current flows from each electrode to its neutral, heating the water between them as it goes. Since the boiler is fully flooded, it is impossible to vary the water level so a false water level is created by inserting porcelain sleeves between the electrodes and neutral shields. These sleeves are raised or lowered (to expose more or less of the electrodes) by a reversible motor mounted on top of the boiler. The effect of this is to reduce, or increase, the flow of current around each electrode. This in turn controls the amount of heat being produced. Convection currents within the boiler ensure the water is kept at an even temperature throughout the vessel.

As heat is generated within the water itself by the passage of the electric current, no part of the boiler is hotter than the water, consequently no thermal problems arise within the boiler. The efficiency of this type of boiler can be in excess of 96%. If for any reason the water level should fall below the electrodes, no load can be taken, making this type of boiler inherently fail-safe.

Electrode hot water boilers are used extensively for space heating, often in conjunction with overnight thermal storage systems on off peak tariffs. Their compact size means that they can be sited in small areas within a building or even on the roof.

It is apparent from looking at the Figures 2.29 and 2.30 that the requirements for safety on these installations is going to be quite stringent. BS 7671 covers the installation of such devices in Section 554. Regulations 554-03-01 to 554-05-04 are particularly relevant to water heaters.

Now would be an ideal point to read through these regulations before continuing with this section.

Try this
Using BS 7671 list the particular requirements of electrode boilers.

Temperature control

Having considered the various methods of heating using electric current we must consider the control of the heating process. We cannot allow heaters to run unchecked as this would give rise to an unacceptable and in some instances a dangerous situation.

The most common method of temperature control is by the use of a thermostat. This may be an integral part of the heater as in the case of fan heaters, night storage heaters, oil filled radiators and the like. It may also be required as a remote sensor for frost protection systems and heating systems other than those powered electrically, such as gas fired boilers.

The operation of a thermostat generally relies on a change in a material when it is subjected to a temperature change. The most simple examples being those of the bimetal strip and gas filled bulb. Let's just refresh our memory as to their operation beginning with the bimetal strip.

Bimetal strip

In Figure 2.31 we can see a simplified construction of a bimetal strip which comprises two materials which have different rates of expansion. These two materials are joined and fixed together along their length. If heat is applied to these materials the one with the greater expansion rate will expand more for any given rise in temperature. If the two materials were not joined the result would be as shown in Figure 2.32. By fixing the materials together the expansion causes the combined strip to bend as shown in Figure 2.33.

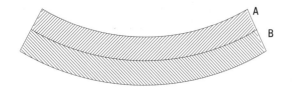

Material A - low rate of expansion

Material B - high rate of expansion

Figure 2.31   Bimetal strip

Figure 2.32   Heat applied to an unbonded strip

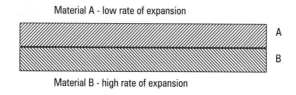

Figure 2.33   Heat applied to a bonded strip

If we position a control switch so that it is operated when the bimetal strip bends we can control the source of heat via this switch. If we make the positioning of the switch operating arm variable to the position of the strip we can adjust the switch to allow the strip to bend greater or lesser distances before the switch operates. This will allow us to vary the temperature change needed to cause the operation of the switch.

A typical room temperature control thermostat is shown in Figure 2.34. A similar device can be used to control a storage radiator. You will see that a small magnet and armature have been added to the assembly. This is to ensure that the operation of the switch is carried out quickly enough to prevent arcing between the contacts being prolonged.

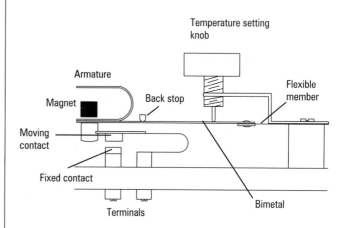

Figure 2.34   Typical bimetal thermostat

A similar result may be achieved using a material such as mercury in a glass phial with electrical contacts. The mercury will expand as the temperature rises and therefore occupy more of the space within the phial. Once it reaches the second contact the circuit is completed and control devices will operate. By varying the angle of the tube we can change the amount of expansion required to operate the switch and thus the temperature change required. This type of device typically operates at extra low voltage.

When we need to control temperature in an area where it is not appropriate to install cables and switching devices there are a number of choices available.

The gas bulb

This forms the basis of the oven thermostat and is used to control mechanical plant and equipment as well as electric heat sources. The basic construction is shown in Figure 2.35 and comprises a gas-filled bulb, a capillary tube and a bellows mechanism. As the bulb is heated, the gas expands through conduction of heat via the bulb. This expansion forces the gas up the capillary tube and expands the bellows. By locating a switch adjacent to the bellows we can use this expansion to operate the switch relative to temperature in much the same way as we did with the bimetallic strip.

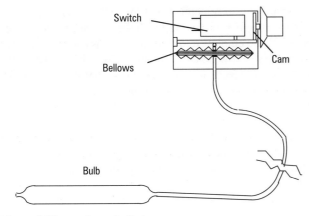

Figure 2.35   A gas bulb thermostat

The rod thermostat

This is the most common form of control for immersion heaters and a typical construction is shown in Figure 2.36. The brass tube and a rod, usually made of invar which has a very low expansion rate, are joined at the tip only. The expansion rate for brass is much higher than that of the invar rod and so as the temperature rises the brass tube expands faster than the rod, effectively pulling the rod into the tube. By locating the switch mechanism so that it is operated by the movement of the rod we can use the device to control the input to the water heater and thus the water temperature.

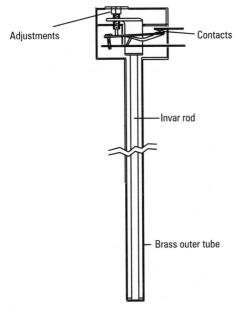

Figure 2.36   Rod thermostat

Thermo-couple

This comprises a junction between two dissimilar metals which, when heated, produce a small e.m.f. Whilst this is normally in the region of millivolts it is sufficient to be monitored by an electronic device or a sensitive voltmeter, scaled to read temperature as opposed to voltage. We can use these to illustrate temperature or, with the addition of the appropriate electronic/relay circuitry, control heating loads.

Variations on these themes are used for all types of heating control and the introduction of heat sensitive electronic components has made the solid state control of shower temperatures and so on commonplace.

These devices are often linked to sophisticated programmed controllers to give a high degree of control within fine tolerances. The savings that can be achieved by the use of such controllers in building management systems can often recover the original outlay within a very short period of time.

Fusible link

In many heaters, storage types in particular, a fusible thermal link was used. This operated in very much the same way as a fuse insomuch as it forms a part of the electrical heating circuit. Should the temperature within the heater rise beyond the melting temperature the link would melt, like a fuse wire, and disconnect the element from the supply. These have largely been replaced within electrical heaters as there were obvious hazards involved with this process.

These devices are still used to control the supply to other types of heating systems, such as gas and oil, where they form part of a cable system which needs to be complete to hold open the fuel supply valve. Fusing of the link will result in the shut off of power. A typical arrangement is shown in Figure 2.37.

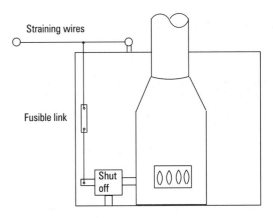

Figure 2.37   Fusible link

Control equipment

Figure 2.38 shows a typical space heating system, including the provision of hot water from the same boiler.

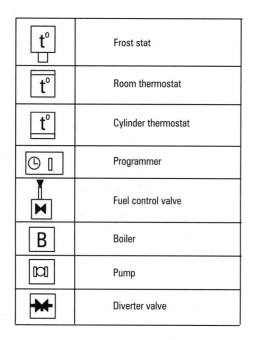

t°	Frost stat
t°	Room thermostat
t°	Cylinder thermostat
⊕ ▯	Programmer
▶◀	Fuel control valve
B	Boiler
▯◁▷▯	Pump
▶▶◀	Diverter valve

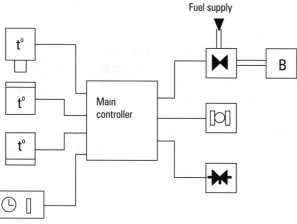

Figure 2.38   Typical control equipment block diagram for space heating system

Heat gain and heat loss

Having considered electric heating methods, we shall now give some thought to their application and use. The purpose of electric heating is to raise the temperature of a particular medium, air, water, oil etc. to a particular level, the GAIN process. The gain process in many cases is linked to a time factor, we don't want to have a heating system that comes on yet does not reach the required temperature for several hours.

Once we have achieved this temperature level we then need to maintain temperature against the changes occurring due to usage. Air loss through opening doors and windows and through the fabric of the building or use of hot water, the LOSS process.

It is important to remember that the planning and design of heating schemes is a complex process. Specialist design engineers are employed to carry out this task and much of the calculation is now done by computer. We shall consider the principles involved and some basic calculations. We will not be carrying out full design and the reasons for this will become apparent as we discuss the topic further.

The first consideration we must make is the amount of heat "energy" that we are going to use to carry out the "work" of raising the temperature of a body such as water or air by a defined amount.

The unit of energy is the Joule and the amount of heat, in Joules, needed to raise the temperature of 1kg of a material by one degree centigrade is known as the "specific heat capacity" of the material. For example water has a specific heat capacity of 4180 joules per kilogram per °C. This means that we require 4180 joules of energy to raise the temperature of 1 kg of water by 1 °C. The symbol for the specific heat of a material is "c" and the amount of heat energy required to produce a given rise in temperature can be calculated by the formula

energy = mass × specific heat capacity × temperature rise

$$\text{joules} = \text{kg} \times \text{J/kgC} \times \text{°C}$$
$$W = mc\,(t_2 - t_1)$$

Water heating

Let's consider the simple storage water heater and look at the process involved.

How much heat is required to raise the temperature of 15 litres of water from 10 °C to 60 °C?

Remember that 1 litre of water has a mass of 1kg.

$$W = mc\,(t_2 - t_1)$$
$$W = 15 \times 4180 \times (60 - 10)$$
$$W = 3{,}135{,}000 \text{ J or } 3.135 \text{ MJ}$$

Now if the heater is rated at 2kW we can determine the time taken for this rise in temperature to take place. Assuming the process is 100% efficient so that there are no losses to be taken into account, the energy required is 3.135 MJ so the time taken, using the formula

$$\text{Time t (seconds)} = \frac{\text{Energy W (Joules)}}{\text{power P (watts)}}$$

$$t = \frac{3135000}{2000}$$

$$t = 1567.5 \text{ seconds or 26 minutes}$$
(to the nearest minute)

If the operation was only 80% efficient the result would be that the energy required would be higher and as a result the time taken to heat the water would increase. So at 80% the input of energy will be

$$3135\,000 \times \frac{100}{80} \quad = \quad 3{,}918{,}750 \text{ joules}$$

so the time taken will be

$$t \quad = \quad \frac{3918750}{2000} \quad = 1959.375 \text{ s or 33 mins}$$
$$\text{(to the nearest minute)}$$

a time increase of 7 minutes.

If we are required to raise temperature within a given period of time, say in the previous example the water needed to be ready within 30 minutes of switching on, then the efficiency of the operation is very important. The only course open to us in our example would be to increase the rating of the heater to reach temperature in the given time, unless we can improve the efficiency of the operation.

One of the problems with heating water in a tank is the losses through the tank into the air. If we can reduce these, we can improve the efficiency both of the heating operation and the storage of the water when it is hot. A good thermal insulator is therefore beneficial to the running cost of the heating system.

Consider the situation where we are to install a water heater system that comprises an immersion heater and a copper tank with a capacity of 60 litres. We know the tank is to be insulated and the overall efficiency of the operation will be 87%. The heater is to raise the temperature of the water from 5 °C to 60 °C in a time of 1.5 hours. What will be the kW rating of the heater we must install?

$$W \quad = \quad mc\,(t_2 - t_1)$$

$$W \quad = \quad 60\,\text{kg} \times 4180\,\text{J}/\text{kg}/^\circ\text{C} \times (60-5)\,^\circ\text{C}$$

$$W \quad = \quad 13{,}794{,}000 \text{ J or } 13.794 \text{ MJ}$$

At 87% efficiency the heat supplied will be

$$\text{Heat supplied} \quad = \quad \frac{\text{heat required}}{\text{efficiency}}$$

$$= \quad \frac{13.794}{0.87}$$

$$= \quad 15.9 \text{ MJ}$$

$$\text{time} \quad t \quad = \quad \frac{\text{Joules}}{\text{Watts}} \quad \text{and so}$$

$$W \quad = \quad \frac{15{,}900{,}000}{90 \times 60}$$

$$= \quad 2944.4 \text{ W} \qquad \text{or } 2.944 \text{ kW}$$

so we would probably fit a 3kW heater.

The water in the tank is now at 60 °C and we assume that the air surrounding the tank is at an ambient temperature of 22 °C. We know that heat will "flow" from a hot body to a cold body and so heat will now radiate from the tank in an attempt to raise the air temperature to that of the water. This will result in the water in the tank "losing heat" into the surroundings.

The insulation applied to the tank will reduce the heat loss from the water and the thermal efficiency of the insulation can be calculated from the formula

Thermal efficiency =

$$\frac{\text{Dissipation from bare metal} - \text{dissipation from insulated metal}}{\text{dissipation from bare metal}} \times 100\%$$

Try this
Calculate the rating of an immersion heater element to be installed in a tank containing 100 litres of water. The water is to be heated from 10 °C to 60 °C over a period of $1\frac{3}{4}$ hours and the efficiency of the operation is 88%. 1 litre of water has a mass of 1 kg, the specific heat of water is 4.2 kJ/kg °C and 1 kWh = 3.6 MJ.

Space heating

If we now look at the similar problems related to space heating, we can see that one factor we must consider is the heat loss from the building. This will have a direct effect on the heat input required and this, in turn, will affect the size of the heaters and the cost of running the system.

All building fabrics will transfer heat through their structure and heavy or dense materials, such as brick and concrete, transfer heat quite readily. These then do not provide good insulation qualities.

Lightweight or low density materials, such as fibre glass and thermal blocks, do not readily transfer heat and therefore provide better insulation properties. For this reason lightweight blocks and insulation materials are favoured in many construction projects. This is because a thermally efficient building offers a considerable saving on running cost over its lifetime.

We must also consider any total air change requirements for areas. In commercial premises it is not uncommon to have air changes per hour given to enable heat calculations to be carried out accurately. Any fully air conditioned environment will have this particular requirement and certain areas will have a mandatory number of air changes required, dependent on the activity and conditions.

Our first step when we are to calculate the heat loss for a building is to consider the various surface areas and the cubic capacity of each room, and note those which are subject to statutory or recommended air changes. Heat loss will be dependent on the surface areas, composition and location of each part of the building structure. Heat will transfer through walls, floors, ceilings, windows, doors and roof spaces at every location where there is a temperature difference between one side of the material and the other.

Ground floors have a heat loss downwards but upper floors will have a heat gain from the heated rooms below. Conversely the top floors will have a heat loss through the unheated roof space whilst the lower floors will be partially insulated by the rooms above. The effect of air changes will depend on the cubic capacity of the room, whether there is a requirement for controlled air changes and to some extent the number of windows and doors, as these opening and closing will cause heat loss.

In order to calculate the heat loss for a particular area we will need to determine the rate at which it will lose heat, to do this we use the "U" value for the material. "U", the thermal conductivity of the material, is found (this is expressed in W/m degree C) and the thickness of the material, in metres, is divided by this value to produce the thermal resistance figure, in m^2 °C/W. This is done for each material and the thermal resistance of each material is added together. So for a single brick wall, plastered on one side and cement skimmed on the other the thermal resistance for each material is added

together. We then take the reciprocal of the total to obtain the overall "U" value for the wall in W/m^2 °C.

In practice the thermal conductivity values for materials are usually produced in table form, for ease of reference, and copies of these are commercially available.

In order to calculate the heat loss due to air change we must calculate the mass of the air change. If we lose warm air then this is replaced by cool air and we must raise the temperature of this cool air to that of the room.

The mass flow rate

$$= \frac{\text{volume} \times \text{density} \times \text{air change rate per hour}}{360}$$

mass flow rate $=$ m^3 × kg/m^3 × l/s

Heat loss

$=$ mass flow rate × specific heat × temperature rise

heat loss $= $ kg/s × J/kg °C × °C

heat loss $= $ J/s $= $ Watts

The loss of heat through the material is

Heat loss $= $ area × U value × temperature rise

W $= $ m^2 × W/m^2 °C × °C

So let's have a look at a typical space heating problem.

We are to establish the heat requirement for a room with the following dimensions and requirements. The air is to be changed twice each hour and the inside temperature is to be maintained at 20 °C when the outside temperature is –2 °C.

Thermal conductivities:

Brick	1.9	W/m^2/°C
Plaster	1.05	W/m^2/°C
Insulating material	0.034	W/m^2/°C
Wooden door	2.84	W/m^2/°C
Window (4 mm glass)	3.1	W/m^2/°C
Concrete	1.44	W/m^2/°C
Asphalt	0.6	W/m^2/°C

Thermal resistances:

Outside surface	0.053	m^2 °C/W
Air gap	0.176	m^2 °C/W
Inside surface	0.123	m^2 °C/W

The density of air is 1.2 kg/m^3 and the specific heat of air is 1.01 kJ/kg °C.

The roof is a concrete slab 150 mm thick with a layer of 25 mm of asphalt on the outside with 15 mm of plaster skim applied to 35 mm of insulating material attached directly to the slab on the underside. The walls comprise two bricks (115 mm of brick) with a 50 mm cavity construction with 15 mm of plaster to the inside.

The room is 15 m × 10.5 m × 3.5 m high and has a window area of 4 m² with a wooden door area of 1.6 m² and a thickness of 50 mm.

In order to determine the heating requirements we shall carry out the calculation in stages to establish the heat loss due to air changes and the loss due to the building fabric. From these we shall determine the total heat requirement.

First the heat loss due to air movement and for this we need to find the total mass of air and the rate of flow. We then determine the heat loss using this value with the specific heat and temperature change, so

The mass flow rate

$$= \frac{\text{volume} \times \text{density} \times \text{air change rate per hour}}{3600}$$

$$\text{mass flow rate} = m^3 \times kg/m^3 \times l/s$$

The mass flow rate

$$= \frac{15 \times 10.5 \times 3.5 \times 1.2 \times 2}{3600}$$

$$= 0.3675 \text{ kg/s}$$

Heat loss

$$= \text{mass flow rate} \times \text{specific heat} \times \text{temperature rise}$$

$$\text{heat loss} = 0.3675 \times 1010 \times 22$$

$$= 8,165.85 \text{ Watts}$$

Heat loss due to air changes = 8.166kW

To calculate the heat loss through the walls, door, window and roof we must calculate the U values for each

$$\text{U value} = \frac{1}{\text{thermal resistance}}$$

Window the U value for the windows is obtained by dividing the thermal conductivity into the thickness of each component and adding these to the inside and outside resistances.

If the glass is 4 mm thick with a thermal conductivity in the order of 3.1 the thermal resistance will be

$$\frac{0.004}{3.1} = 0.00129.$$

U value:

$$= \frac{1}{\text{thermal resistance} + \text{inside resistance} + \text{outside resistance}}$$

$$U = \frac{1}{0.00129 + 0.053 + 0.123}$$

$$= \frac{1}{0.17729}$$

$$= 5.64 \text{ W/m}^2 \text{ }^\circ\text{C}$$

Walls the U value for the walls is obtained by dividing the thermal conductivity into the thickness, for each component, and adding these to the inside and outside resistances in the same way as for the windows.

Plaster:

$$\text{resistance} = \frac{\text{thickness}}{\text{conductivity}}$$

$$= \frac{0.015}{1.05}$$

$$= 0.014 \text{ W/m}^2 \text{ }^\circ\text{C}$$

Brick:

$$\text{resistance} = \frac{\text{thickness}}{\text{conductivity}}$$

$$= \frac{0.115}{1.9}$$

$$= 0.0605 \text{ W/m}^2 \text{ }^\circ\text{C}$$

as this is a double wall total value

$$= 0.0605 \times 2$$

$$= 0.121 \text{ W/m}^2 \text{ }^\circ\text{C}$$

Air gap:

this is given as 0.176 W/m² °C

Total resistance = sum of resistances for plaster, brick, air gap, inside resistance and outside resistance

$$\text{total} = 0.123 + 0.014 + 0.121 + 0.176 + 0.053$$
$$= 0.487 \text{ m}^2 \text{ }^\circ\text{C/W}$$

U value $= \dfrac{1}{0.487}$

$= 2.053 \text{ W/m}^2 \,^\circ\text{C}$

Door

$\text{resistance} = \dfrac{\text{thickness}}{\text{conductivity}}$

$= \dfrac{0.05}{2.84}$

$= 0.0176 \text{ W/m}^2 \,^\circ\text{C}$

Total resistance
= inside resistance + door resistance + outside resistance

total $= 0.123 + 0.0176 + 0.053$
$= 0.1936 \text{ m}^2 \,^\circ\text{C/W}$

U value $= \dfrac{1}{0.1936}$

$= 5.165 \text{ W/m}^2 \,^\circ\text{C}$

Roof

Resistances:

concrete $= \dfrac{0.15}{1.44}$

$= 0.1042 \text{ m}^2 \,^\circ\text{C/W}$

asphalt $= \dfrac{0.025}{0.6}$

$= 0.0416 \text{ m}^2 \,^\circ\text{C/W}$

insulation $= \dfrac{0.035}{0.034}$

$= 1.0294 \text{ m}^2 \,^\circ\text{C/W}$

plaster $= \dfrac{0.015}{1.05}$

$= 0.0143 \text{ m}^2 \,^\circ\text{C/W}$

total $= 0.123 + 0.1042 + 0.0416 + 1.0294 + 0.0413$
$+ 0.053$
$= 1.3655 \text{ m}^2 \,^\circ\text{C/W}$

U value $= \dfrac{1}{1.366}$

$= 0.732 \text{ W/m}^2 \,^\circ\text{C}$

We can now calculate the heat losses for the surfaces from the formula:

$$W = \text{area} \times U \times \text{temperature rise}$$

Glass heat loss $= 4.0 \times 5.64 \times 22$
$= 496.32 \text{ watts}$

Door heat loss $= 1.6 \times 5.165 \times 22$
$= 181.808 \text{ watts}$

Roof heat loss $= 157.5 \times 0.732 \times 22$
$= 2,536.4 \text{ watts}$

Wall heat loss

wall area $=$ area walls $-$ (area door + area window)

$$= [2 \times (15 + 10.5) \times 3.5 - (4.0 + 1.6)] \times 2.053 \times 22$$

Wall heat loss
$= (178.5 - 5.6) \times 2.053 \times 22$
$= 172.9 \times 2.053 \times 22$
$= 7809.2 \text{ watts}$

total heat loss =
losses due to air change + roof + walls + door + window

loss
$= 8166 + 496.32 + 181.808 + 2536.4 + 7809.2$
$= 19189.728 \text{ Watts or } 19.19 \text{ kW}$

Try this

Using the thermal conductivities given below calculate the U values for a brick wall comprising the following: 15 mm plaster, two 100 mm bricks with 50 mm polystyrene insulation installed between them.

Brick	1.9	$W/m^2/°C$
Plaster	1.05	$W/m^2/°C$
Expanded polystyrene	0.034	$W/m^2/°C$
Wooden door	2.84	$W/m^2/°C$
Window	3.1	$W/m^2/°C$
Concrete floor	1.13	$W/m^2/°C$
Asphalt	0.6	$W/m^2/°C$

Wiring for heating systems

The electrical arrangement for total space heating using electric heater(s) must be installed in accordance with BS 7671. If we are using individual heaters these will be supplied on separate radial circuits, large warm air units will usually be supplied on three phases.

The Public Electricity Supplier will provide control equipment to allow the system to operate using off peak energy. That is the necessary switching and metering to facilitate the off peak. If the connected load is not too high the Public Electricity Supplier's control equipment will be adequate to switch the load. If the capacity of their control equipment is exceeded then a contactor will need to be installed. This will switch the load with the operating coil controlled by the Public Electricity Supplier's control equipment.

Figure 2.39 shows a typical layout for "Economy Seven" metering for a domestic installation.

For water heating circuits it is not uncommon to provide a boost control system which allows the consumer to boost the water temperature during the normal consumption periods. This will involve supplying the water heater with both on and off peak connections and a typical schematic is shown in Figure 2.40. A single heating element is supplied via a boost control, which operates in much the same way as a time delay lighting switch, to allow short periods of boost heat to be applied.

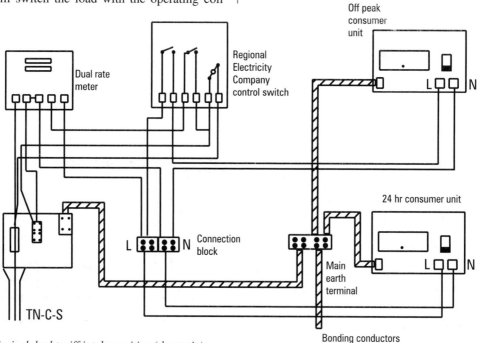

Figure 2.39   *Typical dual tariff intake position (domestic)*

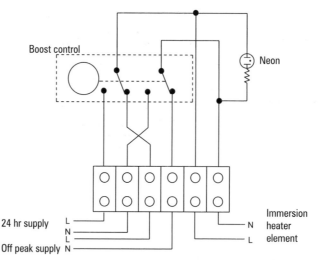

Figure 2.40   *Schematic diagram showing internal wiring of Economy 7 boost control*

We must also consider the control of heating systems powered by other fuels such as gas and oil. These systems are generally a light electrical load which is used to control the operation of the boiler and monitor temperatures in various parts of the system. In general these systems will incorporate some or all the control devices listed below:

- fuel supply control valve
- boiler ignition
- boiler water temperature
- air temperature
- water temperature
- outside air temperature (frost stat)
- water diverting valves
- heating times
- over temperature

The more sophisticated the control system, the more devices we will find incorporated. A typical domestic control circuit is shown in Figure 2.41.

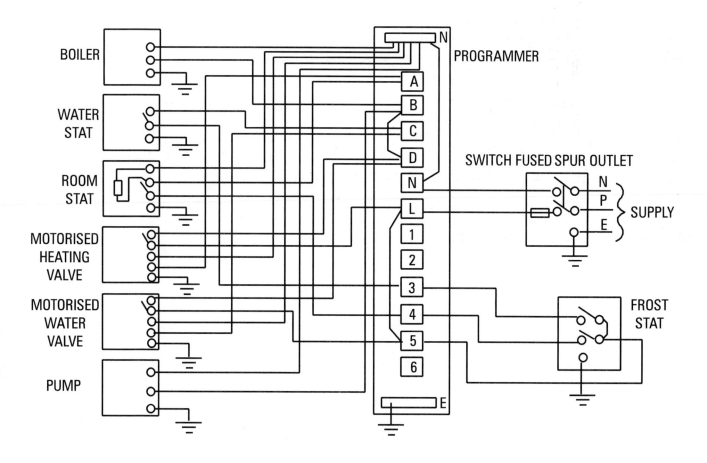

Figure 2.41   Typical wiring for domestic heating controls
            (For clarity the protective conductors have been omitted)

Having now completed Chapter 2 of the module you should consider parts 2a and 2b of the project.

3

Cable Selection

In order to select cables for any application there are a number of details that we need to consider. Although you may have covered many aspects of this before it is as well that we go through the basic requirements as a revision exercise.

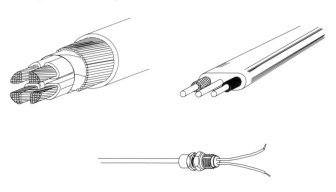

Figure 3.1

On completion of this chapter you should be able to:

◆ list the influences, both external and electrical, which affect the selection of cables
◆ identify the characteristics of the supply required from the electricity supplier
◆ establish maximum demand and apply diversity
◆ calculate single and three-phase loads from given information
◆ calculate load current and from this determine minimum current carrying capacity for live conductors
◆ select protective devices
◆ apply correction factors
◆ select live conductors
◆ calculate voltage drops for given circuits
◆ calculate prospective earth fault current from given data
◆ establish disconnection time of devices under earth fault conditions
◆ check circuits for compliance with shock protection requirements
◆ check for compliance with thermal constraints
◆ calculate minimum size of circuit protective conductors

First we'll look at the details relevant to the electrical system and then the external or non electrical considerations.

Electrical considerations

The design of any electrical installation needs to start with the supply. The requirements for this are determined by the Electricity Supply Regulations 1988 and these put an obligation on the supplier to give the consumer certain information. For example, as we found in Chapter 1, Regulation 30 requires that the supplier must declare to the consumer:

• the number and rotation of phases
• the frequency and
• the voltage.

It goes on to state that under normal circumstances the frequency should be 50 Hz and the supply voltage 230 volts between phase and neutral.

Permitted variations of these values are limited to:
• not exceeding one per cent above or below the declared frequency, and
• 10% above, 6% below the declared voltage.

Also under the Supply Regulations the supplier should provide a written statement of
• the maximum earth loop impedance of the earth fault path outside the consumer's installation (Z_e) and
• the type and rating of the supplier's fusible cut-out or switching device nearest to the supply terminals (I_n).

When all of the information has been established the supplier will issue a written statement similar to that on the following page.

The Supplier

Proposed supply to :   15 Torbayn Close, New Town East

The following information is supplied to meet the requirements of the Electricity Supply Regulations 1988 and Chapter 31 of BS 7671. This information should be made available to the designer of the electrical installation.

i	Nominal voltage	230 volts single-phase
ii	Nature of current and frequency	AC 50 Hz
iii	Prospective short circuit available at the supplier's terminals	16 kA
iv	Earth fault loop impedance (Z_e) of the supplier's network	0.35 ohms
v	Maximum demand permitted to be drawn from the supply	22 kVA
vi	Type and rating (I_n) of the supplier's overcurrent protective device	BS 1361 100 amperes
vii	System arrangement	TN-C-S

Consumer's maximum demand

We need to establish the maximum demand that will be required for the installation.

For small installations this may be done by the use of actual connected load values at an early stage in the design. Larger installations are often assessed on a rule of thumb basis in the initial stages. This is because values such as the actual loads of machines will not be known until later. Designers use a method of approximation based on the use of the building and floor areas involved.

In any event the use of the building determines the level of illumination, temperature and other factors such as air conditioning. All of these can have a direct bearing on the total loading.

Typical rule of thumb values are:

Type of load	Electrical loading/m² of building floor area
Shop lighting	30 W/m²
Small power	10 W/m²
General heating	50 W/m²
kVA for building	Total kW/0.8

External considerations

The external factors that we must consider are contained in Chapter 52 of BS 7671 and include:

- the environmental conditions that will prevail during occupation
- the purpose for which the building will be used on completion
- the type of building construction and the manner in which it is erected
- any special conditions which may exist

We shall consider each of these in turn and the factors which will be relevant in each case beginning with:

Environmental conditions

Temperature – high or low temperatures will have an effect on most modern materials used in cable production. The temperature will also have an effect on enclosures and equipment and therefore we must select these accordingly.

Humidity – having given consideration to the effects of the temperature that may exist we must also make allowance for humidity, i.e. the moisture content of the air.

Water – by this we mean is there likely to be water present in the area. In general this can be viewed in terms of water droplets and the direction in which these droplets are travelling.

Foreign bodies – we can look at this in two stages, firstly as the amount of dust and debris that will be present in the air. Secondly, and this applies generally to enclosures, the likelihood of objects such as tools, fingers or materials being inserted whilst the equipment is live.

Corrosive or polluting substances – in any area where corrosive or polluting substances are likely to be present we must take particular account of the materials used for the installation and their suitability for such an environment.

Mechanical impact – there are many situations where cables are installed and are likely to become damaged due to mechanical impact. We must ensure that suitable protection is provided in such instances.

Vibration – where a wiring system is connected to a machine there is likely to be some vibration and an allowance must be made for this when selecting the method of connection to the fixed wiring system.

Flora or mould growth – this is similar to the consideration that we gave to corrosive substances as many plants and moulds produce corrosive chemicals. Plants may also attract wildlife with the risk of physical damage as well as the increased chemical risk from urine and faeces.

Solar radiation – if we intend carrying out an installation where some part is exposed to the effects of sunlight then we must ensure that the system of wiring and the materials used are suitable for such exposure.

Lightning – if part of the system is located outside of a building then we must consider the possibility of a lightning strike to the system and offer the necessary protection to prevent this from creating a hazard to both the system and the user.

Electromagnetic effects – this covers a number of areas but they are all concerned with the effects of stray electric currents or electromagnetic radiation including electrostatics.

In addition to these considerations we must also have some regard for the effects of natural phenomena such as wind and ground movement.

We must also bear in mind that in certain cases more than one of these conditions will apply at a single location.

Try this

A supply is to be installed to a separate building and the route takes the supply cable through a farmyard, fixed to a brick wall. List the environmental conditions which will apply to the supply cable on this section of the route.

As you can see this is quite a list of points to be considered but we have not quite finished yet. We must also take into account the building's probable use.

IEC Publication 364 contains classification and coding for external influences. Each influence is coded by using a group of two capital letters and a number. The first letter relates to the general category of influence

A: ENVIRONMENT,
B: UTILISATION and
C: CONSTRUCTION OF BUILDINGS

The second letter relates to the nature of the external influence and the number relates to the class within each external influence.

The classification details of external influences are contained in BS 7671 Appendix 5.

A list of External Influences is contained in Guidance Note 1 to the 16th Edition Wiring Regulations, Selection and Erection, published by the IEE.

Try this

By reference to Appendix 5, BS 7671 or Guidance Note 1 establish the external influence referred to under the codes:

AD4

AG3

AE2

AA3

BA2

CA2

CB3

Use of the IP Code (BSEN 60529:1992)

The IP code is one method of checking to see if equipment is suitable for any particular area of risk. A piece of equipment is coded by the use of two numbers. If we consider a length of bare copper overhead line the IP Code would be IP00 as no protection against solid objects or liquids is given.

The first of the two numbers, (numerals 0 to 6, or letter X), refers to the degree of protection offered against solid objects. The higher the index number the greater the degree of protection offered. The second number of the two, (numerals 0 to 8, or letter X), refers to the degree of protection offered against the ingress of liquids. Once again the higher the number the greater the degree of protection.

A third number may be shown and where it is it refers to the degree of protection against mechanical impact.

Some equipment may be coded by use of picture representation. Refer to the IP Code BS EN 60529:1992.

Utilisation of the building

We must consider such factors as:
- the capability of the people using the building
- the density of occupation and
- the probability of people coming into contact with earth potential within the building

Examples of different types of building use could be:

Ordinary domestic/office premises

Factories, workshops and laboratories

Schools, hospitals, children's homes

Power stations and gas installations

Places of assembly – churches, halls, theatres etc.

Remember

Remember that the degree of protection provided by an enclosure is indicated by two numerals. These may be followed by an optional additional letter and/or 1/2 supplementary letters.

i.e. IP23CH

Further details can be found in IEE Guidance Note 1 "Selection and Erection".

Try this

You could find a copy of BS EN 60529 or IEE Guidance Note 1 helpful when answering these questions.

1. An enclosure has a coding of IP 65. Give two examples of situations where this would be appropriate.

2. A remote stop button is to be installed in an area where there is a high level of humidity and is easily accessible to the touch with people working in close proximity. What is the minimum IP Code number that would be suitable for the stop button in these circumstances?

Type of building construction

Our main areas of concern for the building structure are as follows:

Is the material combustible?

Is the shape and size of the building such that it will allow a rapid spread of fire?

Is the building of considerable length or built on unstable ground which could result in the movement of one part of the building with respect to the other?

Does the building contain movable partitions, false ceilings or is the structure completely flexible such as a tent or marquee?

Try this

List 5 different building constructions that have a direct effect on the choice of wiring system.

1.

2.

3.

4.

5.

Special areas for consideration

Hazardous areas

Special consideration should always be given to hazardous areas where there is the possibility that an explosion or flashover could occur. This includes areas for the manufacture and storage of any substance that gives off a flammable vapour or dust. Areas used for the production, storage and sale of petroleum products, paints and sugar manufacturing are typical examples.

Some of the less obvious, at first glance, are sewage treatment and storage areas (particularly farms), grain stores and flour mills. It is vital that a suitable system is installed in these areas to ensure that any spark that may be produced in the normal course of operation (such as the operation of a switch) is contained within the equipment and not released into the explosive atmosphere. This is usually achieved by the use of special enclosures and fittings.

Try this
Using a copy of BS 5345 define the hazardous zones:
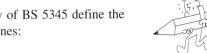

Zone 0:

Zone 1:

Zone 2:

Areas outside these zones are defined as non-hazardous.

An example of a zone 1 area is that within a petrol pump housing. All electrical equipment within that area must be capable of operating in a zone 1 area.

Special installations or locations

BS 7671 has identified a number of installations that require special consideration. These are all found in Part 6 and include areas such as bathrooms, caravans, agricultural and horticultural premises, swimming pools and highway installations.

More details regarding installations requiring special consideration can be found in Chapter 4.

Temporary installations

There is no reduction in the standards for temporary installations although different systems and voltages are often used. The provision of "temporary" lighting on construction sites is often at 110 V to ensure safety as is the provision of small power. In general terms a system of lighting, usually of the festoon type, is provided to give a general level of lighting. This is often below the level that would be expected for a building or area ready for occupation.

Remember

The external factors that affect our choice of system are those external to the electrical installation. These include:

- the type of building construction and the manner in which it is erected
- the purpose for which the building will be used on completion
- the environmental conditions that will prevail during occupation and any special conditions which may exist
- the effects of the surrounding environment on the installation such as wind, building movement and vibration

And a good deal of common sense should be used when taking account of such phenomena.

Having considered the factors that affect our choice of system, albeit briefly, we can begin the process of selecting the conductors we are going to use. The first step will be to select the live conductors.

Conductor selection

Live conductors

We may find that the conditions in which the installation is to be constructed and placed into use cause us to reconsider our cable sizing. As a result of the prevailing conditions we may need to select an alternative installation method or wiring system in the interests of practicality, economy or ease of installation. The first part of the process is to determine the design current of the circuit. Remember this is the current that the load will require under normal conditions.

Design current (I_b)

Let us consider the assumed maximum demand of an installation where every piece of equipment installed is switched on. For this exercise we shall use a simple domestic installation consisting of the following:

	Total
1 × cooker rated at 55 A (no socket outlet)	55 A
2 × ring final circuits each rated at 30 A	60 A
1 × instantaneous shower unit rated at 7 kW	30 A
3 × lighting circuits each rated at 5 A	15 A
	160 A

Details of diversity application are contained in Guidance Note 1 to BS 7671. This is based on a reasonable assumption that not all equipment will be used on full load at the same time.

So by applying diversity we find

Cooker:

> 10 A + 30% of the remaining

Ring circuits:

> 100% of the largest circuit

> + 40% of every other circuit

Shower unit:

> 100% (no diversity for one heater)

Lighting:

> 66% of total current demand

Applying diversity to the above installation

Cooker	10 A
plus 30% of 45 A	13.5 A
Ring circuits	30 A
plus 40% of 30 A	12 A
Shower unit	30 A
Lighting 66% of 15 A	9.9 A
	105.4 A

Diversity may be applied to many types of installation but care must always be taken so that in normal use circuits do not become overloaded.

Try this

A domestic load consists of:

1 × cooker rated at 45 A (no socket outlet)
2 × ring final circuits each rated at 32 A
1 × radial circuit rated at 20 A
1 × immersion heater rated at 15 A
2 × lighting circuits each rated at 6 A

Calculate the total assumed current demand of this installation.

Remember

The diversity allowance must be determined from data, knowledge and experience.

Guesses can cause fires!

Having examined the information that we need to obtain in order to begin the selection process we will do a quick refresher in calculating the current requirements of given loads.

Example

A circuit is to supply a shower unit at 230 V 50Hz and this is rated at 7 kW. What will be the current drawn from the supply if the power factor is unity (1)?

$$\text{power (watts)} = \text{voltage (volts)} \times \text{current (amps)} \times \text{power factor}$$

$$\text{current} = \frac{7000}{230 \times 1}$$

$$= 30.44 \text{ amperes}$$

This is the "Design Current" for this unit and is given the symbol I_b. It is the current that will be drawn by this unit under normal operating conditions.

If we now apply the same principle to a balanced three-phase load of, let's say, 15 kW at 400 V, 50 Hz and a power factor of unity (1) then we have:

$$\text{power} = \sqrt{3} \times U_L \times I_L \times \text{power factor}$$

$$15000 = \sqrt{3} \times 400 \times I_L \times 1$$

so the line current will be

$$I_L = \frac{15000}{\sqrt{3} \times 400 \times 1}$$

$$= 21.65 \text{ amperes}$$

So for this balanced three-phase load the design current I_b will be 21.65 amperes per phase.

Let's now go back to a single-phase example. Taking a domestic 3 kW electric heater supplied with 230 V 50 Hz, the current is

$$\text{Required current} = \frac{3000}{230 \times 1}$$

$$= 13.04 \text{ amps}$$

Remember that being a resistive load the power factor is taken as unity (1).

So if 13.04 amps is the current required for the heater to operate normally then logically the fuse or circuit breaker used to protect the circuit must be capable of carrying this current without damage or deterioration for an indefinite time.

Try this

1. A domestic water heater is rated at 2.8 kW. Calculate the current demand when it is connected to a single-phase 230 V supply.

2. An industrial water heater is rated at 18 kW. Calculate the current demand when it is connected to a 400 V three-phase supply.

Selection of protective device

The actual rating will depend on the type of device used so we must decide on the type before we can select the rating. Some of the types and ratings are shown in Table 3.1.

In the case of our 3kW heater we are going to use a BS 1361 type fuse and by reference to the manufacturer's data, or the table in Table 3.1, we can see that the nearest size of fuse is a 15A.

Remember

$$I_n \geq I_b$$

This rule must always be applied so for example if the design current of a circuit was 20.5 A we could not use a 20 A BS1361 fuse for protection and we have to go up to a 30 A device.

So for our domestic heater we know that we need to install a 15A or 16A protection device. First we need to establish the minimum current carrying capacity of the cable required to supply the load. The current carrying capacity of cables are considered in BS 7671 as I_z and I_t where:

I_z is the current carrying capacity of a cable for continuous service, under particular conditions

and

I_t is the current tabulated in Appendix 4 of BS 7671 for the type of cable and method of installation applicable to a single circuit in an ambient temperature of 30 °C.

For the purpose of selecting our cable, we generally calculate the cable current carrying capacity to make the selection from the tables in Appendix 4 of BS 7671. As we have seen these values are based upon a set of criteria which may not apply in every case. For example, it is quite common for cables to be run grouped with other cables, through areas with an ambient temperature greater or less than 30 °C or under other conditions which will affect the current carrying capacity. In order to determine the cable size required we calculate the minimum I_t value by applying factors to compensate for any variations to the conditions applied in the Appendix 4 tables. The current carrying capacity can then be used to determined cable size directly from the tables in Appendix 4.

Remember: We calculate the minimum value of current carrying capacity for the cable required by applying factors to compensate for any changes in the prescribed conditions for Appendix 4 of BS 7671. By doing this the cable size can then be found directly from the tables in the Appendix 4.

Protection Type	Current Rating
BS EN 60269-1:1994 (BS 88)	2 4 6 10 16 20 25 32 40 50 63 80 100 125 160 200 250 315 355
BS EN 60269-1:1994 (BS 88)	355 400 450 500 560 630 710 800
BS EN 60269-1:1994 (BS 88)	2 4 6 10 16 20 25 32
BS1361	5 15 20 30 40 45 50 60
MCB BS EN 60898BS3871 Single pole Double pole Triple pole	Types (2 & 3 BS 3871) B & C 6 10 16 20 32 40 50 63

Table 3.1

Before we can determine the minimum current carrying capacity we must consider the conditions which apply to the particular cable we are installing. That is the method of installation and the conditions under which the cable will be operating.

Method of installation

Whatever type of material we use for a conductor it will have some resistance. We know that the factors which affect the value of this resistance are

- the material from which the conductor is made
- the cross sectional area (c.s.a.) of the conductor
- the length of the conductor
- the temperature of the conductor

From the above list the only one which we can really affect is the temperature of the conductor. We must try to keep the heat produced in the conductor to a minimum during operation. To do this we need to consider the type of location and conditions and how these will affect the amount of heat that a cable can dissipate (give off).

The main factors which affect this ability are:

- ambient temperature
- grouping
- thermal insulation
- type of protection device used

Ambient temperature

This is the temperature of the surroundings of the cable. It is often the temperature of the room or building in which the cable is installed. Now, the hotter the surroundings the less heat the cable will be able to give off, (we put food in a warm oven to keep it hot). If the surrounding temperature is low then the heat given off will be greater and so the cable would "run cooler", and so could give off more heat.

Grouping

If a number of cables are run together, they will all produce heat when they are carrying current. The effect of this is that they reduce the heat dissipation of each other. The same effect is used by groups of animals huddling together to keep warm. If we keep the cables separated then this effect will be minimised.

Thermal insulation

This has a similar effect to wrapping a cable in a fur coat on a hot summer's day and the heat produced cannot escape. *Regulation 523-04 and Table 52A* in the Regulations give derating factors for cables surrounded by thermal insulation. For cables in contact with thermal insulation on one side then Reference Method 4, in Appendix 4 of BS 7671, makes allowance for this condition.

Examples of reference methods

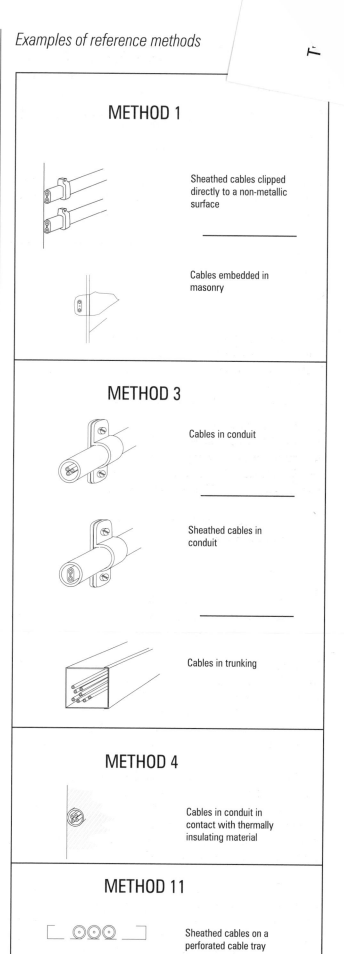

METHOD 1

Sheathed cables clipped directly to a non-metallic surface

Cables embedded in masonry

METHOD 3

Cables in conduit

Sheathed cables in conduit

Cables in trunking

METHOD 4

Cables in conduit in contact with thermally insulating material

METHOD 11

Sheathed cables on a perforated cable tray

Table 3.2

Type of protection device used

The devices we use to protect a cable generally operate when too much current passes through them. This excess current will produce more heat within the cable and unless the device can disconnect quickly enough damage may be caused to the cable insulation. Under extreme conditions the insulation may catch fire. The time in which the device should disconnect depends upon the purpose for which the circuit is to be used and the likely risk of electric shock to the user of the installation. In general terms there are two principal disconnection times based upon the provision of shock protection against indirect contact. Those which apply to the majority of installations are:

- 5 seconds for circuits supplying fixed or stationary equipment
- 0.4 seconds for circuits supplying socket outlets which may have hand held equipment plugged into them.

There are however certain types of installation locations, such as agricultural installations, and particular uses which require different disconnection times. Section 413 and Part 6 of BS 7671 should be consulted for the disconnection times required for protection against indirect contact in such locations.

Determining the factors

Now that heat has been identified as the problem we can try to find ways of avoiding it. Whilst we cannot prevent some conditions, we can take some precautionary steps. Often another route can be found or cables can be spaced so there is no contact between them.

As it is not always possible to avoid these problems a system of factors can be used to make sure the conductor is large enough to cope with them.

Generally correction factors are found in *Appendix 4* of BS 7671 but on occasions it is necessary to look elsewhere.

Ambient temperature

The tables used for cable selection will be based on a particular ambient temperature. Generally, as in the case of BS 7671, Appendix 4, this is 30 °C. Any cables installed in an ambient temperature above 30 °C will need an adjustment to their rating as they will be unable to give off as much heat.

A set of factors for these conditions are given in BS 7671, Table 4C1 and 4C2. Complete Table 3.3 using a copy of BS 7671. You will notice that the type of insulation used also has an effect on the factors given in the tables. The factor related to ambient temperature is referred to as C_a.

Because of the characteristics of the BS 3036, semi-enclosed fuse, a separate table must be used where these are used as the means of protection. Complete Table 3.4 using a copy of BS 7671.

Grouping

Factors related to the grouping of cables are contained in BS 7671 in Table 4B1. Complete Table 3.5 to show values for grouping factors. You will see that the method of installation also has some bearing on the factor to be used. It is important to remember that these factors are applied to the number of circuits or multicore cables that are grouped and not the number of conductors. When adding circuits to an existing conduit or trunking it is important to remember that the design of the existing circuits may be affected by the addition. The factor related to the grouping of cables is known as C_g.

Try this

You will need to refer to a current copy of BS7671 to be able to enter the appropriate factors in the following tables. Note the number of the table in BS 7671 from where you obtained the information.

Table 3.5

Correction factors for grouped cables:

Reference method of installation	Correction factor (C_g)							
	Number of circuits or multicore cables							
	2	3	4	5	6	7	8	9
Enclosed (3 or 4)								
Single layer multicore on cable tray (11) T								
S								

T – touching

S – spaced, "spaced" means a clearance between adjacent surfaces of at least one cable diameter (De).

The factors for Mineral Insulated Cables are given in a separate table in BS 7671.

Try this

A general purpose PVC cable is run through an ambient temperature of 50 °C and is protected by a 10 A fuse to BS 88. Explain what effect replacing the fuse with one to BS 3036 may have.

Thermal insulation

The coming into contact of cables with thermal insulation is considered as two separate scenarios.

Totally surrounded: If the cable is surrounded by thermal insulation then, if the length of run is up to 400 mm the derating factors, relevant to the length, are given in Table 52 A in BS 7671. If the length of run exceeds 500 mm then the derating factor which must be applied is 0.5 and this factor is referred to as C_i.

Contact on one side only: If the cable is in contact with thermal insulation on one side only the situation does not require a derating factor to be applied. However when we are selecting using the tables in BS 7671, Appendix 4, we must use Installation Method 4.

Type of protective device used

The disconnection time of most protection devices can be predetermined from reliable characteristics. With the exception of the BS3036 fuse, all are capable of isolating faults in the required times. If we select the BS3036 device then we must ALWAYS use a factor of 0.725 when calculating current carrying capacity of the cable it is protecting. This value is derived from BS 7671, Regulation 433-02-03, which states that, for this type of device I_n should not exceed 0.725 × the lowest cable current carrying capacity. The factor related to the use of the BS 3036 fuse is referred to as C_f.

For BS 3036 fuses
$$I_n \leq (0.725 \times \text{lowest cable current carrying capacity in the circuit protected})$$

Remember

Correction Factors

C_a ambient temperature

C_g grouping

C_i thermal insulation

C_t operating temperature of conductor

Applying factors

The purpose of applying the factors is to ensure that the conductor is large enough to carry the current without excessive heat being generated. Once the best route and installation method have been selected, if the cable is affected by adverse conditions which impair the heat dissipation we must take alternative action. It is important to ensure that the best route and installation method have been selected because the only alternative left to resolve the situation is to increase the size of the cable.

When a protective device operates it generally relies on an overcurrent to do so. If a device is rated at 20 A then this is the current that it will carry for an indefinite period without deterioration. It follows then that the device will only register current beyond its rated value as overcurrent. We must therefore use the rating of the device, I_b, as the starting point for our calculations to determine the minimum value of I_t.

If more than one of the conditions which require the application of derating factors exists, then we must consider the worst case scenario. If, as an example, a cable runs through an area of high ambient temperature, then it is grouped with several other cables at another location and is finally run totally enclosed in thermal insulation at another point then we consider all the factors and apply the most onerous. If however more than one condition applies at a single location then we must apply all those factors relevant to that location.

To establish the minimum value of I_t for any circuit we must divide the current rating of the protective device by all the factors. If any factor does not apply we give it the value 1.

So
$$I_t \geq \frac{I_n}{C_a \times C_i \times C_f}$$

where C_f is the factor for a BS3036 fuse.

If no factors need to be applied then

$$I_t \geq \frac{I_n}{1 \times 1 \times 1}$$

and in this case

$$I_t \geq I_n$$

Try this

List 3 advantages of changing BS3036 semi-enclosed fuses for BS EN 60898 Type 2 mcbs.

1.

2.

3.

Example

A PVC/PVC multicore cable is protected by a 16 A fuse to BS88 and is run through an area where the ambient temperature is 40 °C. Determine the minimum value of I_t.

As there is only one factor to apply

$$I_t \geq \frac{16}{C_a}$$

So we turn to Table 4C1 (remember we are using a BS88 fuse). That means we have the value of 0.87 for general purpose PVC.

So our value for I_t will be

$$I_t \geq \frac{16}{0.87} \geq 18.39 \text{ A}$$

You can see that the current carrying capacity has increased to over 18 A as a result of the effect of a higher ambient temperature.

We will assume that the same cable is to be run through the 40 °C temperature and is also grouped with 3 other cables. If it is also protected by a BS 3036 fuse then we must apply all these factors to determine the minimum I_t. So we have;

$$I_t \geq \frac{16 \text{ A}}{C_a \times C_g \times 0.725}$$
$$C_a = 0.94 \text{ from } Table \ 4C2$$

When we come to establish the value of C_g using Table 4B1 of BS 7671, we find that we need yet more information about the circuit that we are to install. This time we need to know the method of installation. This is the method of wiring, so we need to know how the cable is to be installed. For the normal domestic installation the method would be to run the cables clipped directly to the surface of the building so that is the method that we shall use.

$$C_g = 0.65 \text{ from } Table \ 4B1$$

We must remember to use the total number of cables that are bunched together. For this example this is the 3 other cables plus the 1 cable that we are to install giving a total of 4.

So to complete the calculation we have

$$I_t \geq \frac{16}{0.94 \times 0.65 \times 0.725}$$
$$I_t \geq 36.119 \text{ A}$$

As you can see, this has a considerable effect on the current carrying capacity of the conductor required. In practice it would be a more sensible idea to run the additional cable separate from the others and change the protective device. This would then reduce the factors that need to be applied.

Try this

A cable is to be installed in an area with an ambient temperature of 45 °C at one stage in its run. In a separate area the same cable is run with 3 other cables. For the last part of the cable run it goes through a wall which has 500 mm of thermal insulation. At which stage of the cable run does the most onerous situation exist and what correction factors need to be applied?

Try this

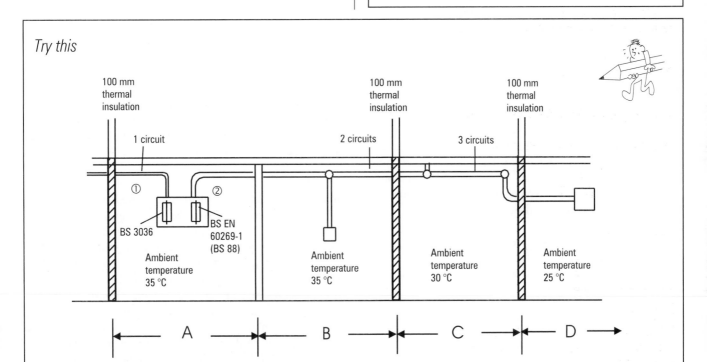

Circuit ① a PVC/PVC cable clipped direct to the surface.
Circuit ② a PVC cable run in conduit fixed direct to the surface.

Determine the most onerous location and the correction factor(s) which will need to be applied to determine the minimum value of I_t for;

(a) Circuit ①

(b) Circuit ②

Selecting the live conductors

The first stage in determining the minimum cross sectional area of conductor required for any circuit is to identify the maximum load it is expected to carry. Once the load is determined, a protection device can be selected. So that the minimum size and most cost effective cable can be used, a BS3036 fuse should be avoided.

Once the protection device is selected the cable run must be examined in detail, as we have already seen, and questions asked such as:

- What is the maximum ambient temperature?
- How many other circuits are adjacent?
- Is there any thermal insulation that should be considered?
- What is the installation Reference Method?

When these are all answered, the cable run must be checked to see how many factors apply at the same time and which is the "worst" condition.

Example

A 4.5 kW single-phase load connected to a 230 V 50 Hz supply has a power factor of 0.8. The cable supplying the motor is to be installed in conduit with one other circuit of similar load and is 15 m from the distribution board. The ambient temperature is not expected to exceed 30 °C and no thermal insulation is involved. The motor is to be protected with a BS88 type fuse with the appropriate characteristics.

First calculate the total load current.

As this is an a.c. induction load with a power factor, the total kVA must be calculated first.

$$\text{As } P \quad = \quad U I \cos \theta$$

$$U I \quad = \quad \frac{P}{\cos \theta} \quad \text{or}$$

$$kVA \quad = \quad \frac{kW}{\text{power factor}}$$

$$= \quad \frac{4.5}{0.8}$$

$$= \quad 5.625 \text{ kVA}$$

$$\text{Current} \quad = \quad \frac{kVA}{V}$$

$$= \quad \frac{5.625 \times 1000}{230}$$

$$= \quad 24.46 \text{ A}$$

Next determine a suitable rating of protection device.

The nearest above 24.46 A in a BS 88 is a 25 A so

$$I_n \quad = \quad 25 \text{ A}$$

Before a cable can be determined, the factors that may affect the cable size must be taken into account.

$$
\begin{aligned}
C_a &= 1 \text{ (ambient temperature 30 °C)} \\
C_i &= 1 \text{ (no thermal insulation)} \\
C_f &= 1 \text{ (no BS3036 fuse)} \\
C_g &= 0.8 \text{ (two circuits)}
\end{aligned}
$$

Calculate the minimum value of I_t.

$$I_t \quad \geq \quad \frac{I_n}{C_a \times C_i \times C_f \times C_g}$$

$$\geq \quad \frac{25}{1 \times 1 \times 1 \times 0.8}$$

$$\geq \quad 31.25 \text{ A}$$

This means that the cable must be rated at least 31.25 A. From BS 7671, Table 4D1A (single core PVC insulated cables) columns 1 and 4 (Reference method 3, and 2 cables single-phase), a 4 mm^2 cable is rated at 32 A.

This shows that the cable is capable of carrying the current but that is not the end of the calculation. As we have already seen, all conductors have a resistance which generates heat when current flows. This resistance will also cause a voltage drop along the cable, which could result in the inefficient operation of the load. It will also cause a voltage drop in the cables resulting in the inefficient operation of the motor. Voltage drop is not restricted to the cables within an installation. The electricity supplier will also have problems with voltage drop along their cables. The Supply Regulations limit by law the supplier's voltage drop to 6% of the declared voltage. This means for a 230 V single-phase supply the voltage could be

$$
\begin{aligned}
230 \times 6\% &= 13.8 \text{ V} \\
230 - 13.8 &= 216.2 \text{ V}
\end{aligned}
$$

BS 7671 allows a maximum voltage drop within an installation of 4% of the declared voltage. So within an installation this could be

$$
\begin{aligned}
230 \times 4\% &= 9.2 \text{ V} \\
230 - 9.2 &= 220.8 \text{ V}
\end{aligned}
$$

Looking at the worst possible case the voltage on a load in an installation could be

$$230 - 13.8 - 9.2 = \quad 207 \text{ V}$$

This should be looked at in relation to the power that equipment can output. A 1 kW resistive load on 230 V has a resistance of

(Remember $P = \dfrac{U^2}{R}$ so $R = \dfrac{U^2}{P}$)

$$R = \frac{230^2}{1000}$$

$$= 52.9\ \Omega$$

The resistance is relatively constant so we can assume that even when the voltage is reduced to 207 V it will still be about the same. This means that the power output of the load when supplied with 207 V will be

$$P = \frac{U^2}{R}$$

$$= \frac{207^2}{52.9}$$

$$= 810\ W$$

So although the voltage has dropped by 10% (6% or 4%) the power output has reduced by almost 20%. This shows how important it is to keep the voltage drop in an installation to a minimum.

It should also be remembered that the final circuit in an installation is often only part of the voltage drop chain. The 4% referred to in BS 7671 is the total from the main intake to the furthest point on any circuit.

Going back to our 4.5 kW load, we shall assume that there are other voltage drops in sub-main cables and the maximum that there can be in the final circuit cable is 4.5 V.

Now we need to calculate what the voltage drop would be if we used the 4 mm² we calculated. To do this we need to go to BS 7671, Table 4D1B, which shows the voltage drops associated with the cables shown in BS 7671 Table 4D1A. From Columns 1 and 3 it can be seen that a 4mm² cable has a voltage drop of 11 mV/A/m. From this we can determine the actual voltage drop in the cable when it is carrying the full load of 24.46 A.

$$\text{Voltage drop} = \frac{11 \times 24.46 \times 15}{1000}$$

$$= 4.0359\ V$$

As this is less than 4.5 V the 4 mm² cable is suitable.

Having completed the process of selecting the live conductors, we must now consider the selection of the protective conductors. In order to do this we must first consider the earth fault path.

Protective conductors

Earth fault path

The impedance of the earth fault path, known as Z_s, plays an important part in the system as it will regulate the amount of current that flows in the earth fault path.

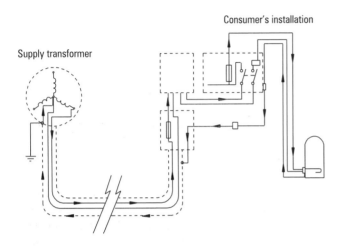

Figure 3.2   *TN-S system*
Earth faults are returned to the supply
transformer through the metal sheath of the
supply cables

If we look at the circuit diagram in Figure 3.2 we can see the case of the electric immersion heater connected to the circuit protective conductor. This is in turn connected to the consumer's earth terminal and then via the supply system to the star point of the transformer. This is best shown on the TN-S system for clarity. All the parts of the circuit which are the consumer's responsibility, and connected to the consumer's earth terminal, are "exposed conductive parts". These are all part of the electrical installation and include conduit, trunking, all circuit protective conductors and cases of appliances and equipment.

Remember that in the case of a fault to earth all the "exposed conductive parts" of the installation become live for the period of time that it takes for the protective device to disconnect the circuit from the supply.

The earth fault current will flow around the earth fault loop as shown in Figure 3.2.

This path will offer an impedance to the flow of the fault current. This impedance will be the sum of the individual impedances of the conductors which go to make up the earth fault loop. As we can see, the loop comprises the transformer winding, the phase conductor of the supply and the consumer's circuit up to the fault. From the fault to the consumer's earth terminal is the circuit protective conductor and from the consumer's earth terminal back to the star point is dependent on the type of system. For our calculations we assume that the fault itself offers no resistance to the flow of current.

In order for us to determine the current that will flow in the event of a fault we need to establish the impedance of this earth fault loop. If the installation is already installed then the earth fault loop can be measured using a line earth loop impedance tester. If we are designing a system then we need to calculate the value of earth fault loop. To do this we use the value of earth fault loop impedance of the supply system which we obtained from the electricity supplier. This value is known as Z_e.

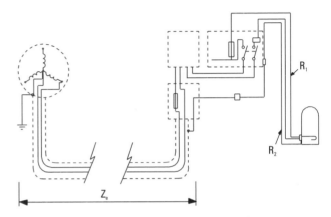

Figure 3.3   *TN-S system showing Z_e, R_1 and R_2*

To this we must add the impedance of the part of the loop that is made up by the circuit conductors. This will be the phase conductor up to the fault and the circuit protective conductor back to the consumer's earth terminal. These values are known as R_1 and R_2 respectively and are taken to the point on the circuit furthest from the supply to establish the worst case, i.e. when the conductor impedance is at its maximum. Figure 3.3 illustrates Z_e and $R_1 + R_2$ for our immersion heater circuit.

We can best see how this is done by using our earlier example of the 4.5 kW inductive load and calculating the earth fault loop impedance.

Conductor resistances

From earlier calculations we can assume we have established that the size of the live conductors we are using are 4.0 mm^2. The standard sizes of composite cables are supplied with the cpc one size smaller than the live conductors, so it would be a good idea to take this as a starting point for determining the value of the earth fault loop impedance. We know then that the size of the cpc is going to be 2.5 mm^2 and so we are ready to calculate the value of the earth fault loop impedance of the consumer's part of the system (R_1 and R_2).

To carry out the calculation we need to know the following details
- the length of the circuit conductors
- the cross-sectional area of the phase conductors
- the cross-sectional area of the circuit protective conductor (this is often the same as that of the phase conductor but not always)
- the impedance of the phase and protective conductor per metre

This last detail we get from tables, so let's take a look at the section of these as shown in Tables 3.6A and 3.6B. (Derived from manufacturers' data)

Table 3.6A

Cross-sectional area	Resistance (milliohms/metre)	
Phase conductor	Plain copper	Steel
1	18.10	
1.5	12.10	10.7
2.5	7.41	9.1
4	4.61	7.5
6	3.08	6.8

To allow for temperature rise under fault conditions the following correction factors must be applied.

Table 3.6B

	Copper	Steel
Multiplier	1.38	1.2

The important thing to remember is that the values given are in **MILLIOHMS** per metre.

For our calculation we require the resistance of a 4.0 mm^2 phase conductor with a 2.5 mm^2 protective conductor. Working from Table 3.6A the resistance of a 4.0 mm^2 copper conductor is 4.61 milliohms/metre and a 2.5 mm^2 copper cpc is 7.41 milliohms/metre. This gives us a total value of 12.02mΩ/m. This is not the end of the calculation though as we need to know the resistance under fault conditions. This could mean the temperature rising in the conductors and increasing their resistance. To allow for this a factor of 1.38, from Table 3.6B, must be applied to our total value.

We now have all the details that we need to calculate the value of the earth fault path within the consumer's installation, this will be $R_1 + R_2$.

length of run	=	15 m
resistance from Table 3.6A	=	12.02 mΩ/m
multiplier from Table 3.6B	=	1.38

REMEMBER:
Always use the multiplier from Table 3.6B.

The total value will be calculated using the formula:

$$resistance = \frac{m\Omega/\,m \times length \times multiplier}{1000}$$

$$resistance = \frac{12.02 \times 15 \times 1.38}{1000}$$

$$= 0.249\ \Omega$$

Try this
A cable is 30 m long with live conductors of 6 mm^2 and a cpc of 4 mm^2 all of copper. Calculate the $R_1 + R_2$ value for this cable under fault conditions.

Earth fault loop impedance value

To establish the total value of earth fault loop impedance we must add to this the earth fault loop impedance of the supply, Z_e. We shall assume this is a TN-C-S system and the electricity supplier has quoted a value of 0.35 ohms. So using a Z_e of 0.35 ohms we get a total earth fault loop impedance of

$$Z_s = Z_e + (R_1 + R_2)$$

$$= 0.35 + 0.249$$

$$= 0.599\ \Omega$$

If this had been a TN-S system the electricity supplier would be more likely to quote 0.8 Ω for Z_e.

This would give a total earth fault loop impedance of

$$Z_s = Z_e + (R_1 + R_2)$$

$$= 0.8 + 0.249$$

$$= 1.049\ \Omega$$

Try this
Referring to the previous "Try this" if the Z_e value is 0.35 Ω calculate the total value for Z_s.

So what is the significance of the value of Z_s? If we refer to BS 7671, Tables 41B1, 41B2 and 41D, we find the types and ratings of protective devices listed. Below each rating is given a maximum value for Z_s for the device.

Table 41B1 gives the maximum values for Z_s for fuses, which will provide disconnection within 0.4 seconds with U_0 of 230 V. Remember U_0 is the nominal phase to earth voltage on a TN system. Table 41D gives maximum values for fuses which will provide disconnection within 5 seconds with U_0 of 230 V. Table 41B2 shows the maximum values for mcbs which will provide disconnection within both 0.4 and 5 seconds with U_0 of 230 V.

Values in the BS 7671 Tables relate to the maximum earth fault loop impedance for general circuits with U_0 of 230 V. Section 6 of BS 7671 gives other values for Z_s related to special installations.

We can see that the maximum values for Z_s in BS 7671, Section 4 Tables, are different for each table. The reason for this is that the disconnection times are based upon the risk of electric shock. Portable equipment has an inherent higher shock risk as these items are likely to be held by the user of the installation. Fixed equipment, generally, has a lesser risk of shock to the user as the equipment is not held by the user. For this reason the 0.4 second disconnection time applies to portable equipment and effectively a high current needs to flow in the event of a fault. The 5 second disconnection time applies to fixed equipment generally, and a lower current is required to achieve this disconnection time.

We will assume that on the circuit supplying our load we are using a 25 A BS 88 to supply fixed equipment. We shall therefore use BS 7671, Table 41D, to check for compliance of the circuit. The maximum value of Z_s is given in the table as 2.40 Ω for a 25 A BS88 type fuse. As our value is 0.599 Ω this circuit does comply.

If we apply a simple Ohms law calculation to our circuit then we can find the current that will flow in the event of a fault to earth. This is known as the prospective earth fault current I_f and is found by using the formula

$$I_f = \frac{U_o}{Z_s}$$

where:

I_f is the prospective earth fault current
U_o is the supply voltage and
Z_s is the earth fault loop impedance

so

$$I_f = \frac{230\ V}{0.599\ \Omega}$$

$$= 383.97\ A$$

Now that we know the value of the prospective earth fault current we can further check the compliance of the circuit by using the tables in *Appendix 3* of BS 7671. These tables are the simplified time/current curves for the various types of protective devices.

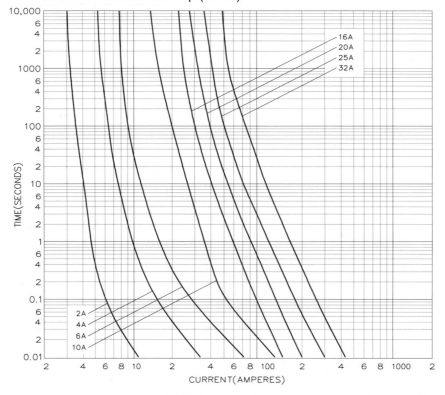

Time-Current Characteristics of Fuse
Bussmann AAO 2-32 Amp (BS88)

Figure 3.4   Time/current characteristics of fuses to BS 88. Reproduced with kind permission of Bussmann Division, Cooper (U.K.) Ltd.

So to further check the compliance of our circuit we can use the value of I_f and the appropriate time/current curve to establish the disconnection time for the device. To do this we use the table for the BS88 fuse (Figure 3.4).

We move along the X axis until we reach the value of I_f for our circuit (384 A), we then move vertically up from this point until we bisect the curve for the 25 A device. As the circuit we are considering supplies fixed equipment the disconnection time must be no more than 5 seconds. The value we get is in fact less than 0.1 seconds so our circuit complies with requirements, but we expected this as the maximum value of Z_s calculated already indicated this to be the case.

Try this

Using the time/current characteristics in Figure 3.4 for a BS88 fuse determine the fault current flowing when a 25 A fuse operates in 3 seconds.

Up to now we have found the size of conductors needed to give compliance with requirements for shock protection. Whilst doing this we established that a high current flows in the event of an earth fault under the correct conditions. In providing shock protection this current must be carried by the circuit protective conductor for the time that it takes for the protective device to operate and disconnect the circuit from the supply. The circuit protective conductor is generally a smaller cross-sectional area than the associated live conductors. As the current carried by the cpc may be quite high, a great deal of heat may be generated whilst the fault current flows.

We must make sure that whilst the fault current is flowing, the heat produced will not cause damage to the conductors or the insulation and material surrounding them. If sufficient current flows for a long enough period of time the heat produced could be such that the insulation catches fire. Once this happens disconnection of the supply will not extinguish the flames.

Obviously this situation cannot be allowed to occur. To prevent it we must ensure that the circuit protective conductor is large enough to carry the fault current, for the period of time needed for the device to disconnect the supply, without excessive heat being produced.

There are thermal constraints placed upon a circuit which are detailed in BS 7671. These ensure that the circuit protective conductor is large enough to carry the earth fault current without detrimental effect to the conductor, the insulation or the surroundings.

In order to find the minimum cross sectional area of the circuit protective conductor we require the following information
- the prospective earth fault current "I_f" Amperes
- the time taken for the protective device to operate with this fault current "t" seconds
- the constant k which is related to the type of circuit protective conductor and its method of installation

The way in which these are related to the minimum size of the circuit protective conductor is by a formula known as the adiabatic equation.

The minimum cross sectional area of the conductor is known as "S" and the formula we use is

$$S = \frac{\sqrt{I^2 \times t}}{k}$$

(The value of I_f is used for this calculation.)

It is important that the calculation is carried out in the correct sequence in order to obtain the correct answer.

Let's consider the circuit supplying the the 4.5 kW inductive load to see whether it meets the requirements for thermal constraints. We know that the prospective earth fault current is 400 A. We also found that disconnection time will be less than 0.1 of a second. Now we need to find the value of the constant k for our circuit. The values for k are found in BS 7671, Tables 54B to 54F. At the head of each table is the description of the method of installation of the circuit protective conductor. For our type of cable and the method of installation we find the value of k from Table 54C to be 115.

Try this

Fill in the appropriate values for k from Table 54C.

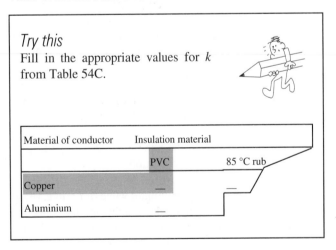

Material of conductor	Insulation material	
	PVC	85 °C rub
Copper	—	—
Aluminium	—	

If we now put these values into the formula we have

$$S = \frac{\sqrt{400^2 \times 0.1}}{115}$$

We must carry out the calculation in the correct sequence to give the correct solution

stage one: Square the value of I_f, i.e. $I_f \times I_f$

stage two: Multiply the result by the time t seconds

stage three: Take the square root of the result

stage four: Divide the answer by the value of k

In our particular case this would be

$$400 \times 400 \quad = \quad 160000$$

$$160000 \times 0.1 \quad = \quad 16000$$

$$\sqrt{16000} \quad = \quad 126.49$$

$$\frac{126.49}{115} \quad = \quad 1.099 \text{ mm}^2$$

The minimum size of copper to give compliance with the thermal constraints is 1.099 mm² and we have installed a 2.5 mm² conductor. We can see from this that the circuit fulfils the requirements for thermal constraints and so our circuit complies.

If the size of cpc installed proves to be insufficient then a larger cross sectional area conductor must be used. This will of course have an effect on the value of the earth fault loop impedance as a larger cpc will reduce the impedance and a higher current will flow. This in turn will reduce the disconnection time.

We could calculate the minimum requirements based on the known criteria that applies to our circuit. For example we know that for a socket outlet circuit the maximum disconnection time is 0.4 seconds. If we know the type and rating of the protective device we can establish the minimum value of I_f to give disconnection in 0.4 seconds. This data and the value of k for the type of cpc and its method of installation will allow us to establish the minimum size of cpc to give compliance under the worst conditions.

To do this we must use the minimum value of I_f to give the disconnection time required, the maximum disconnection time and k for the type of cpc and its appropriate installation method. If we carry out this one calculation we can establish the minimum size of cpc to comply with the absolute worst conditions. We can then ensure that the size selected will be within a usable range. This can be particularly useful when designing circuits for installation in conduit and trunking where the size of cpc can be varied with relative ease. It will also ensure that a great deal of time is not wasted carrying out repeated calculations to establish an acceptable size.

Try this
A circuit supplying 230 V socket outlets is protected by a 32 A fuse to BS88 and has a value of Z_s of 0.76 Ω. Calculate the minimum cross-sectional area of protective copper conductor required if the insulation material is 90 °C thermosetting PVC and the cpc is incorporated within the cable.

PROJECT

Now we have completed Chapter 3 of the module we should consider part 3 of the project.

We shall be involved in the selection of suitable size of cable so we will need a copy of BS 7671 and some manufacturers' details on LSF, SWA, LSF cables.

4

Installations in Areas requiring Special Consideration

During the course of our cable selection process reference was made to installations which have requirements beyond those applicable to general installations. In this chapter we will be considering some of these special installations. The intent is to provide an overview of the requirements for installations involving the following in general terms;

- Areas of flammable vapour and dust
- Agricultural and horticultural
- Corrosion and erosion
- Fire alarms
- Standby supplies
- Lightning protection

and to consider cables for special installations.

On completion of this chapter you should be able to:

- ◆ identify the installation techniques employed in hazardous areas
- ◆ identify the risks and considerations associated with petrol filling station installations
- ◆ identify, from given criteria, areas in which risk of fire or explosion may be present
- ◆ identify areas giving rise to the need for special considerations
- ◆ identify cables suitable for given environmental and hazardous conditions
- ◆ describe methods of protection against corrosion and erosion
- ◆ describe and explain systems of electrically operated fire detection and alarm systems
- ◆ identify the need for standby supplies and identify the means of achieving them
- ◆ identify and describe the means for protection of structures from lightning

BS 7671 deals with special locations in Part 6 and this book has not focused on all those installations covered within Part 6. The Institution of Electrical Engineers produce Guidance Note 7 covering special installations which provides further guidance on those installations considered in Part 6 of BS 7671. Guidance Note 7 also considers areas such as marinas, exhibitions and show stands, gardens and medical locations and therefore we will not cover these items in this chapter. It would be a worthwhile exercise to obtain a copy of Guidance Note 7 and familiarise yourself with the requirements for these special areas. If you are using this study book towards the City & Guilds 'C' course assessments the areas covered in Guidance Note 7 may prove to be beneficial.

The nature of installations in areas of increased risk are such that there are many British and European standards produced detailing the requirements for electrical installations in these hostile and hazardous environments. The appropriate standards should always be consulted when designing installations in potentially dangerous situations.

Our first consideration will be for installations in hazardous areas which include explosive and flammable locations. Electrical installations, by the very nature of their operation, produce heat and switching functions create sparks. If heat or sparking occurs in an atmosphere which contains flammable dust or vapour then an explosion may result.

Figure 4.1   *A petrol filling station is an example of a hazardous area*

Explosive hazardous areas

There are many different examples of situations that may have explosive gas, dust or vapour present at sometime. As the concentrate of flammable material varies, dependant upon the type of activity involved, not all situations represent the same risk of explosion. To help to assess the risk, and therefore the degree of protection that is required, a zone classification system has been adopted.

Zone 0:
> is one in which an explosive gas-air mixture is present continually, or present for long periods.

Zone 1:
> applies to a situation where an explosive gas-air mixture is likely to occur in normal operation.

Zone 2:
> is where the risk is reduced and is applicable where an explosive gas-air mixture is not likely to occur in normal operation and, if it does occur it will only exist for a short time.

Where an area is outside these zones, it is defined as being non-explosive.

Equipment selection

There are a number of ways of protecting equipment for hazardous areas, these are categorised and designated a letter to indicate to which specification it has been designed.

Types of protection:

Ex 'd'   Flameproof enclosure

Ex 'i'   Intrinsically-safe

Ex 'p'   Pressurised Apparatus

Ex 'e'   Increased safety

Ex 'N'   Non-Sparking (Restricted Breathing)

Ex 's'   Special Protection

Ex 'o'   Oil-Immersion

Ex 'q'   Powder/Sand Filling

Ex 'm'   Encapsulation

EEx   is used to show that the apparatus is in conformity with the harmonised European standards

These can be grouped into different methods, as shown in Table 4.1.

Table 4.1

Method		Type of Protection
Designed to prevent the flammable mixture reaching a means of ignition	Ex 'N' Ex 'm' Ex 'p' Ex 'o'	(Restricted Breathing) Encapsulation Pressurisation Oil Immersion
Designed to prevent any ignition from arising	Ex 'e' Ex 'N' Ex 'm'	Increased Safety Non Sparking Encapsulation
Designed to prevent any ignition from spreading	Ex 'd' Ex 'q'	Flameproof Enclosure Powder Filling
Designed to limit the ignition energy of the circuit	Ex 'i'	Intrinsic Safety

Now let's look at each type of protection in more detail.

Ex 'd'

indicates a "Flameproof enclosure" for electrical apparatus. Should flammable gas or vapour enter the enclosure, any explosion that results should not cause damage to the apparatus. It is also a requirement that no flames or hot gases are transmitted to the external explosive atmosphere through any part of the enclosure, which includes joints and any structural openings.

Examples of flameproof enclosures are shown in Figure 4.2.

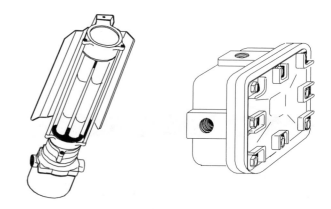

Figure 4.2   Flameproof enclosures

Ex 'p'

denotes a pressurisation system which should have one of the following features. Either

- the system maintains a positive static pressure within the apparatus

or

- the system has a continuous flow of air or inert gas to neutralise or carry away any flammable mixture entering or being formed within the enclosure.

Essential to both systems is the monitoring of the pressurisation system and purging schedules to ensure their reliability.

Table 4.2 shows the minimum action that should take place in the event of a pressure failure.

Table 4.2 Minimum action on pressure failure

Classification of area	Minimum action on pressure failure	
	Enclosure contains apparatus suitable for Zone 2	Enclosure contains ignition capable apparatus
Zone 1	Alarm*	Alarm and de-energise ignition capable components
Zone 2	Not applicable	Alarm*

*When the alarm is operated, immediate action should be taken to restore the integrity of the system.

Ex'e'

indicates "increased safety" where
- electrical apparatus is used that does not produce sparks or arcs and
- the maximum operating temperature does not exceed either the ignition level of the explosive atmosphere or the thermal stability of the materials in use,

it can be considered to offer increased safety.

Components which may cause a spark are normally excluded. The other components are designed to substantially reduce the likelihood of fault conditions which could cause ignition. This is achieved by
- reducing and controlling working temperatures
- ensuring electrical connections are reliable
- increasing the effectiveness of insulation
- reducing the probability of contamination due to the ingress of dirt or moisture

'N'

indicates a type of "non-sparking protection". It can also be shown as IEC'n'. Precautions are taken with connections and wiring to increase reliability, though not to as high a degree as for Ex'e'. Where internal surfaces are hotter than the desired T* rating they can be tightly enclosed to prevent the ready access of flammable mixtures into the internal parts. This is the "restricted breathing enclosure" technique. The employment of this technique means that high ingress protection ratings, IP66 and above, are incorporated in the design.

T* rating - Apparatus may be marked with a "T" symbol which indicates its temperature classification; normally between T1 and T6. Further details on these classifications may be found in Pts 2 & 7 of BS EN 60079 and are covered briefly later in this chapter.

An example of an 'N' type floodlight suitable for Zone 2 hazardous areas is shown in Figure 4.3.

Figure 4.3   'N' type floodlight

's'

is used for equipment with "special protection". This relates to types of electrical apparatus which, by its nature, does not comply with the requirements for established forms of protection. However to achieve the 's' rating it must be shown, by test, that the equipment is suitable for use in the appropriate prescribed zone. As, by definition, this category does not have a set of rules it will not become part of the harmonised EN series of standards.

Ex'o'

Oil-immersion is a technique that has been used for many years primarily for switchgear. The spark is formed under oil and the venting of the resultant byproducts is controlled.

Ex'q'

Powder/sand filling is normally used in a vented enclosure in which components with incendiary potential are mounted. It is primarily of use where the incendiary action is the abnormal release of electrical energy by rupture of fuses or failure of components such as capacitors. Usually it is used for components inside Ex'e' or Ex'N' apparatus.

Ex'm'

"Encapsulation" is a method of enclosing components with incendiary potential so that the components cannot come into contact with flammable atmospheres. The process is used to control the surface temperature and fault conditions.

Ex'i'

shows that equipment or apparatus has been designed and constructed to be "intrinsically-safe". This is a method of protection that is based on restricting the electrical spark energy to below that which will ignite the specific gas mixture. This means that it is not only the electrical apparatus that has to meet this requirement but also all associated interconnecting wiring and equipment. It is not only the energy that has to be controlled but also the heat.

The intrinsically-safe concept is possible only if specially designed equipment is installed to control the working and fault limits. The main "component" is a safety barrier such as shown in Figure 4.4.

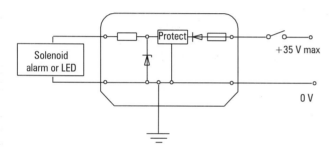

Figure 4.4   Basic circuit of a safety barrier

This safety barrier is installed outside the hazardous area but protects the monitoring equipment inside. Both the supply and monitoring equipment are kept earth free so that no multiple loops can be set up. The safety barrier controls the maximum energy that can be used in the hazardous area by use of a shunt diode, with voltage-regulating circuit.

To meet the reliability requirements for the different zones, European IS (intrinsically-safe) equipment is made and certified in one of two forms, having different degrees of redundancy in the key components. Ex'ia' means that safety is maintained with up to two "countable" component or other faults. This may be used in, or connected into, Zone 0 hazardous areas. Ex'ib' means safety is maintained with up to one countable component or other fault and it may be used or connected into Zone 1 hazardous areas. Certain components, such as wire-wound resistors, are regarded as "infallible" in respect of certain types of fault, as are certain methods of assembly. Manufacturers work closely with the standards and limit the number of components and probable faults, that have to be counted.

To ensure that equipment with suitable protection is installed in the appropriate zones, typical categorisation is as follows-

Table 4.3

Zone	Type of protection	
0	'ia'	intrinsically-safe apparatus or system to this specification
	's'	special protection (specifically certified for zone 0)
1	Any from zone 0 plus	
	'd'	flameproof enclosure
	'ib'	intrinsically-safe apparatus or system
	'p'	pressurization, continuous dilution and pressurised rooms
	'e'	increased safety
	's'	special protection
2	Any from zone 0 or zone1 plus	
	'N'	or IEC 'n' type protection
	'o'	oil-immersion
	'q'	sand-filling

Alternatively, apparatus that in normal operation is not able to produce arcs capable of ignition, sparks or high temperatures may be acceptable in zone 2. However for this to be the case they must have been assessed by persons who should:

• be familiar with the requirements,
• have access to the necessary information,
• utilise similar test apparatus where required.

The "high surface temperatures" referred to are usually shown on the equipment rating plate as a 'T' classification. This identifies the maximum surface temperature that will be reached by that particular item of equipment. Some examples of the 'T' classifications are :

Table 4.4

T class	Maximum surface temperature °C
T1	450
T2	300
T3	200
T4	135
T5	100
T6	85

The type of gas or vapour which gives rise to the hazard is a further factor which needs consideration. Gases have been formed into groups to define the "gas hazard" and some of the categories used are as follows:

Table 4.5

Representative gas	Apparatus group
Methane	1*
Propane	11A
Ethylene	11B
Hydrogen	11C

* For underground mining applications

BS 5501 defines the apparatus and gas groups in more detail.

It has been found that there is no correlation between ease of ignition by hot surfaces and by sparks; the two mechanisms of ignition are entirely different. For example, hydrogen is easily ignited by a low-energy spark of 20 μJ but has a high ignition temperature of 560 °C, whereas acetaldehyde requires a high-energy spark of 150 μJ but has a low ignition temperature of 140 °C.

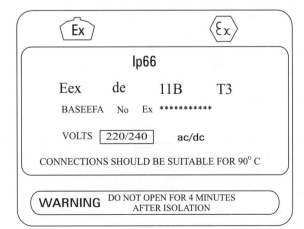

Figure 4.5   Marking plate

The manufacturer's marking plate in Figure 4.5 shows an example of how all the relevant information can be put together.

Having now categorised all of the necessary information, a selection procedure can be adopted. This is usually carried out in the following sequence -
i)   determine the classification of the area
ii)   determine the temperature classification of the hazardous situation
iii) determine the apparatus type of protection required
iv)   take account of the environmental conditions.

Try this
Find an example of a manufacturer's marking plate on a piece of equipment. Copy the plate in the space below and state the type of equipment the plate was attached to.

Pressurised and pipe ventilated systems

There are many situations in which the atmosphere is so dirty and dusty that the operation of the electrical equipment may be seriously impaired.

The dust may be relatively harmless in terms of fire or explosion risk, such as that generated by industrial processes involving stone, sand or clay. In other industries, air-borne dust particles may present a fire or even an explosion risk as previously described.

To prevent dust contamination of items such as control panels, the panel builder may install a fan which draws air through a filter and discharges into the panel enclosure. The small amount of positive pressure thus produced is sufficient to repel any dust particles which may have entered the panel through door joints and so on.

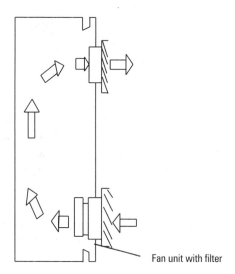

Fan unit with filter

Figure 4.6

Where the atmosphere immediately surrounding the equipment is unsuitable for this purpose, air from outside the building or from a clean adjacent zone can be drawn in through ducts or pipes. This technique is also used for ventilation where the local air quality or temperature is unsuitable.

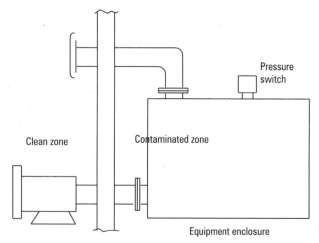

Figure 4.7

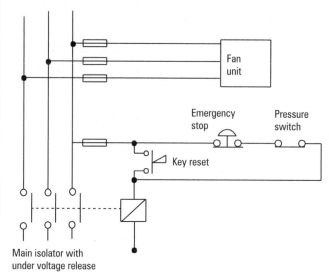

Main isolator with under voltage release

Figure 4.8   *Circuit diagram for a pressurised and pipe ventilated system*

Electrical installation

The process of selecting the appropriate type of installation equipment must take into account many of the factors that have already been discussed. There are a number of general factors that relate to all hazardous areas where an explosion could occur. These general factors should be taken account of before the requirements for specific installation types are considered.

Wiring systems

The wiring supplying specially selected equipment must offer at least the same degree of protection as the equipment itself. Consideration has to be given to the run of cables and enclosures, so that a hazard from one area is not transmitted into another.

Generally electrical apparatus is kept clear of zone 0 areas unless it meets the requirements for intrinsically-safe 'ia'. This also applies to wiring systems even if they are supplying equipment in other zones. Where it is essential to install equipment in Zone 0 mineral insulated cable with the appropriate terminations can be used.

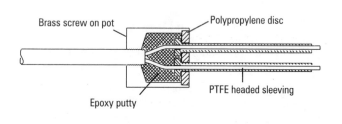

Figure 4.9   *MIMS seal*

The wiring systems that are acceptable in zones 1 and 2 are many and varied. They include:

- cables in steel conduit
- cables protected against mechanical damage, for example thermoplastics or elastomer insulated, screened or armoured cable, with or without a lead sheath and with PVC, csp, pcp, cpe, or similar covering
- seamless aluminium sheath with or without armour
- mineral insulated metal sheathed (Figure 4.9)
- flexible cable or cords with metallic screen or armour with PVC, csp, pcp, cpe or similar covering

---

Remember

csp – **c**hloro**s**ulphonated **p**olyethylene

pcp – **p**oly**c**hloro**p**rene

cpe – **c**hlorinated **p**oly**e**thylene

Where steel conduit is used as a flameproof enclosure 'd' there are a number of special precautions to be adopted. These include

- its use must be kept to a minimum when installed out of doors
- it must not have aluminium or flexible cables installed in it
- running couplers must not be used
- all screw joints must be tight and backed with a lock nut
- stopper boxes must be installed at each point of electrical connection unless the device forms an integral part of the apparatus and is certified
- stopper boxes must be installed at all points where it goes from a hazardous area to a non hazardous area.

Earthing and earth fault protection

In general BS 7671 is used to ensure the earthing meets the necessary safety requirements.

Isolation

Again BS 7671 together with the Electricity at Work Regulations 1989, form the basis of the requirements. There is however a specific requirement in BS 5345 that suggests that at a suitable point outside the hazardous area there should be a single or multiple means of isolation.

Inspection and testing

Both inspection and testing must be considered at the initial stage and at each subsequent stage in the life of the installation. In most circumstances compliance with the requirements of BS 7671 will automatically mean the relevant points are covered. This type of installation does, however, have some special considerations which must be included at the initial and subsequent inspections.

Some of the additional items that need to be considered at the initial inspection are:

- Apparatus appropriate for the area classification
- Correct temperature classification
- Integrity of enclosures
- Cable entries and stoppers and the like are complete and appropriate
- Earthing and bonding meet requirements
- Adequate environmental protection
- No unauthorised modifications

Figure 4.10   Enclosure

When carrying out subsequent inspections there are additional items that need to be considered, some of which are:

- Corrosion of enclosures, fixings etc.
- Damage to apparatus or wiring systems
- Excessive accumulation of dust and dirt
- Loose electrical connections
- Loose fixings, glands, conduit, stoppers etc.
- Conditions of enclosure gaskets and fastenings
- Oil or compound leaks
- Bearing conditions
- Inadvertent contact between rotating and fixed parts
- Integrity of guards
- Ratings and types of lamps
- Vibration
- Correct working of relays and protective devices
- No unauthorised modifications or adjustments
- Maintenance carried out other than to manufacturers' recommendations

Initial and periodic testing

Where test equipment is to be used in a hazardous environment care must always be taken not to create a dangerous situation when testing. The test equipment must therefore be suitable for use in the clarification of zone where the tests are carried out. Three tests for both initial and periodic testing are identified. These are:

i)   Insulation resistance
ii)   Earth electrode resistance (where applicable)
iii) Earth fault loop impedance

In addition to these tests any earth bonding conductors will need checking for continuity.

Potentially explosive installations

There are many installations which encompass some area in which a potentially explosive atmosphere exists. These can include processing plants that involve a high concentrate of dust, such as flour mills or timber workings, or it may be flammable vapours like cellulose or alcohol. Other examples are operating theatres where oxygen and anaesthetics are the problem, or meteorological stations where hydrogen is used to fill weather balloons. In these last two examples sparks from static electricity can be as big a problem as that from electricity supplies and equipment. Here everything has to be bonded together and to earth. This means EVERYTHING – all metal fixtures, window frames, and electrical equipment. Portable steel tables and trolleys are linked together to reduce the possibility of a static discharge.

One example that we all come into contact with is the petrol filling station and we shall look at this particular type of installation in some further detail.

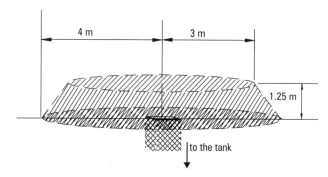

Figure 4.11   *Hazardous area surrounding an underground storage tank*

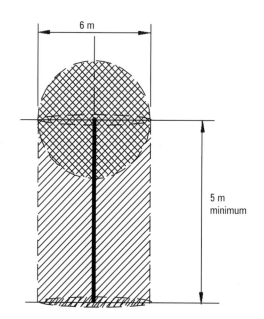

Figure 4.12   *Hazardous area surrounding a vent pipe*

Try this
(BS 7671, Appendix 2)
1. Who may grant licences in respect of premises where petroleum spirit is stored?

2. Identify the statutory regulations to which reference should be made for installations in potentially explosive atmospheres.

i)

ii)

iii)

iv)

The petrol filling station
In this one example all of the hazardous zones apply and electrical equipment is essential to the operation of the site.

Within each petrol filling station there are three areas that have a high concentrate of petrol vapour. These are:
i)   the filling point to the underground storage tanks
ii)   the ventilation pipe to the storage tank
iii) the dispensing pump from the storage tank to the vehicle
Each of these, and the area around them, is classified into zones. These are shown in Figures 4.11 to 4.13.

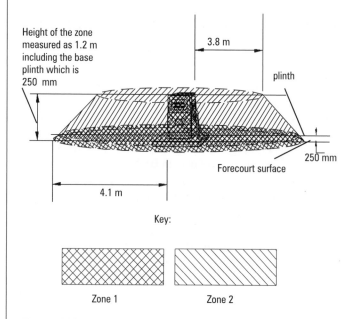

Key:

Zone 1   Zone 2

Figure 4.13   *Hazardous area at a low level metering pump*

These form the basic requirements for the zones but when there are variations the zones have to be adjusted accordingly. An example of this is where there are a number of dispensing pumps placed on an island. Here the zones keep the same proportions but are extended over the longer length, as shown in Figure 4.14.

4.1 m   4.1 m

Figure 4.14

For simplicity the dispensers shown in the diagrams have their delivery hose in the main housing of the unit. In practice there are many variations of this and the zones may need adjusting accordingly. For example some dispensers have the hose coming out at high level and next to the outlet is a sightglass to show fuel is available. In such cases the zone areas are altered to accommodate the particular arrangements. The precise requirements can be obtained from "Guidance for the Design, Construction and Maintenance of Petrol Filling Stations", published by The Association for Petroleum and Explosives Administration & the Institute of Petroleum. If there is still doubt BS 7117 should contain all of the necessary information.

Installation wiring

It is commonplace to have elaborate kiosks and shops on the forecourt to meet customer requirements. The design and position of these is critical, if the construction is such that the unit has an opening within one of the hazardous zones, then the whole of the interior must be designed to meet the requirements for that zone. This means that if a shop door opens inside a zone 2 area, all of the electrical installation inside the shop must be to zone 2 standard. A mistake at this stage can be very difficult and costly to overcome. The installation in the kiosk would normally come under the general requirements of BS 7671 unless it is incorporated into a hazardous zone. The electrical installation along with the rest of the petrol forecourt, must be approved and licensed by the local authority. Under The Petroleum (Consolidation) Act 1928, Local Authorities are empowered for the licensing, enforcement and appeals related to premises where petroleum is stored and/or dispensed. General guidance is provided in "Guidance for the Design, Construction and Maintenance of Petrol Filling Stations".

Petrol forecourt installations usually consist of equipment that is specifically designed for the purpose. The dispensing pumps are an example of design and build to meet the requirements for particular zones of protection. The electrical supply terminates at a point inside the enclosure where the internal wiring has been preconnected. The wiring from the distribution point to the pump has to be carried out on site to meet all of the relevant requirements. Within Zone 1 and Zone 2 areas mineral insulated cable served overall with a PVC sheath or equivalent is preferred. The terminations on these must be of an approved type for the appropriate zone employing an earth tailed pot. The termination should be protected by a suitable shroud to protect it and the cable from corrosion.

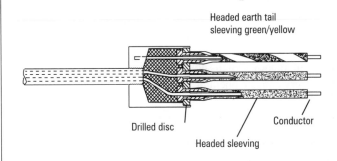

Headed earth tail
sleeving green/yellow

Drilled disc   Conductor

Headed sleeving

Figure 4.15   MI termination with earth tail

An alternative cable can be PVC, or equivalent, insulated, steel wire armoured and PVC, or equivalent, sheath cable. The termination must of course be suitable for the zoning of the hazardous area to maintain the integrity of the enclosure. To ensure the termination provides a sound connection to the enclosure, an earth tag washer with a separate protective conductor must be installed and connected to the earthing terminal within the enclosure.

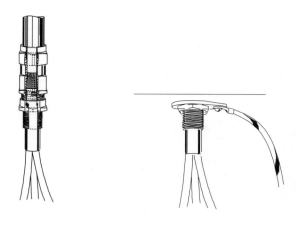

Figure 4.16   Earth plate on SWA

In most cases the cables have to be installed underground or in site-formed ducts. In either case they should be laid at least 0.5 m deep or protected against mechanical damage. Where cables are laid directly in the ground they should be covered by cable covers or suitable marking tape. When any form of ducting is used precautions have to be taken to prevent flammable vapours, or liquids, going from one zone to another or into a non hazardous area.

Dispensing pumps are quite complex pieces of equipment. It is common to find battery back up supplies, time delays, remote monitoring and lighting units, all part of the equipment. This can mean that more than one supply is required to each pump and the isolation is not straightforward. It is a requirement that every pump/dispenser circuit, not intrinsically safe, should be provided with an isolating switch or isolating circuit breaker, for disconnection from the source of electrical energy. In cases where there is more than one source of electrical energy, such as battery support supplies, suitable warning notices must be fitted inside the enclosure and at the point of isolation.

An emergency switch should be provided at
• every console operator position, in self service stations and
• outside the hazardous area, within the forecourt,
and be clearly visible and identifiable to the public.

These switches must disconnect the supplies to all dispensing pumps and their integral lighting. The restoring of the supplies should be such that the reset is located and arranged so that it can only be carried out manually by an authorised person. A notice should be fitted adjacent to each means of operation of the emergency switching device bearing the words **PETROL PUMPS SWITCH OFF HERE**. The means of operation of the switch should be coloured red against a yellow background.

Circuit protection from fault currents is very similar to other installations for short circuit and overload. For pump motors, integral lighting and ancillary circuits they should be protected by suitably rated multiple pole circuit breakers arranged to

break all live conductors including the neutral. Protection against indirect contact should be provided by means of earthed equipotential bonding and automatic disconnection of supply. Alternatively equipment of Class 11 construction can be used where it is under effective supervision in normal use. To provide indirect contact protection all circuits feeding forecourt equipment must have a disconnection time not exceeding 100 mS in the event of a fault developing to earth.

Where the supply to the premises forms a TN-S or TT system the earthing arrangements can be the same as any other installation, but if the supply forms a TN-C-S system other procedures must be adopted. The TN-C-S system uses the neutral conductor as a combined conductor for earth fault current purposes. This arrangement is generally not acceptable for petrol filling station supplies and unless the supplier can give a separate earth connection other arrangements have to be made. This usually means ignoring the earth connection given by the supplier and installing a consumer's earth electrode and effectively making the supply a TT system. In these circumstances an RCD is installed to protect against earth leakage currents.

Other electrical equipment associated with a petrol filling station, such as canopy lights, loudspeaker systems and high voltage neon signs, are usually placed at positions outside any hazardous area. Where this is the case, the installation must comply with BS 7671, but care must always be taken to ensure that vent pipes and road tanker delivery points are not likely to cause a problem.

Guidance for the Design, Construction and Maintenance of Petrol Filling Stations recommends the use of RCBOs for the protection of circuits supplying equipment within the hazardous areas, and that each circuit should be supplied via a separate RCBO. When a single device is used to provide indirect contact protection the device must break phase and neutral conductors, for example multiple pole MCBs or RCBOs, and so fuses may not be used.

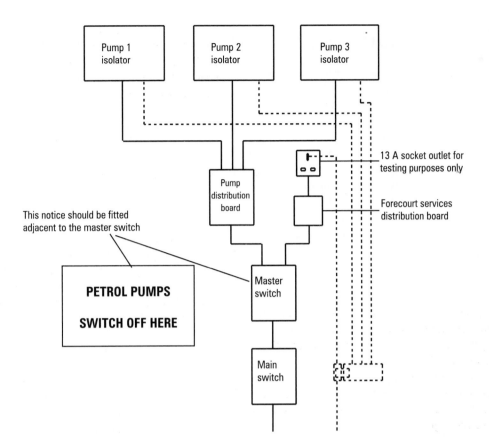

Pump 1
isolator

Pump 2
isolator

Pump 3
isolator

Pump
distribution
board

13 A socket outlet for
testing purposes only

This notice should be fitted
adjacent to the master switch

Forecourt services
distribution board

PETROL PUMPS

SWITCH OFF HERE

Master
switch

Main
switch

Figure 4.17

Inspection and testing

The inspection and testing can be categorised as gas-free or not gas-free. New installations should be inspected and tested whilst still gas-free and this should basically comply with BS 7671. The Guidance for the Design, Construction and Maintenance of Petrol Filling Stations provides programmes for the verification of petrol filling stations. Programmes one and two cover new and major refurbishment projects.

Programme one covers the pre-commissioning verification during construction and pre-commissioning. This programme deals with the inspection and testing of the installation by a competent person whilst the area is gas free. The programme provides details of the inspection and testing requirements for the construction phase of the installation.

Programme two deals with the Initial Verification carried out once the electrical installation is complete. The Guidance also provides model forms for an inventory checklist and initial assessment which should be completed and retained with the electrical documentation. A checklist for the visual inspection is provided within the guidance as are details of the required tests.

Programmes 3 to 6 deal with the periodic inspection and testing of existing installations. Which of these programmes is used is dependant upon two factors:
1. are there site records available? and
2. are the dispensing pumps and equipment certified to BS 7117?

Details of the application of a particular programme are given in the Guidance at Table 14.1.

For further information regarding the requirements of the installation, inspection, testing and additional certification you should refer to the Guidance for the Design, Construction and Maintenance of Petrol Filling Stations.

General hazardous areas

So we have considered the requirements for hazardous areas where there is a risk of explosion or fire. It is important to remember that whilst these risks occur in some obvious locations, such as petrol stations they also arise in some less obvious places. Some typical examples are mines, water and sewage treatment works, flour and grain processing and storage, photographic processing and so on. When carrying out the design or installation in any area which involves the use of chemicals or flammable materials we must establish the nature of the hazard. Having done so we must ensure that a suitable electrical installation is provided, designed and constructed to prevent danger.

There are, of course, areas which may create a hazard to the user of the installation by the nature of the environment or the equipment in use. These areas may not be subject to the precautions required for an area subject to the risk of fire or explosion, but need special consideration in their own right. Such areas include;

- exterior lighting exposed to the environment and vandalism
- street furniture
- construction sites
- swimming pools and saunas
- equipment with high earth leakage currents
- caravans and caravan parks

Whilst this list is not exhaustive it does illustrate some of the more common areas of risk which require special consideration.

Remember

When undertaking design or installation work it is important to consider whether any particular risk may apply. The fact that in the final analysis it may not be applicable does not mean that it should not be considered. Adapting and altering an installation to provide additional protection against a hazard, not considered at the design stage, can be a very costly business.

In order to establish whether there are any special risks associated with the electrical installation there are a number of facts which must be determined. We can consider the requirements due to the environment in broad terms as follows;

Is it?

wet
humid
hot
cold
chemical - corrosive
flammable
toxic
sterile

Are?

livestock present

Does it have?

restriction of movement

What is the likelihood of?

electric shock
burns
static discharges

Having established the nature of the environment and the particular risks involved, we need to establish what regulations, standards and guidance exists relating to the installation. The current versions should then be obtained in order that a full appraisal of the requirements can be undertaken. Once this is completed then the installation design can be completed.

Some of the installations are so specialist that they have their own regulations and staff require special qualifications. Mines and quarries are typical examples. The onerous conditions that electrical installations are required to cope with makes these situations limited to special consideration. So much so that they are generally out of the scope of the BS 7671. These installations are however, covered by the Electricity at Work Regulations 1989. Many of the old sets of Regulations for individual mines and quarries were revoked when the 1989 regulations came into force. As these installations are so specialist work should only be carried out by personnel who are conversant with the special requirements.

There are other types of installation that are hazardous in different ways. Heavy plant maintenance depots have for example a variety of hazards, which include:

- wet
- dirt
- oil
- variations of temperature with weather
- mechanical damage and
- toxic fumes

to mention but a few.

The electrical equipment used in this particular environment is quite varied and includes;

- compressors
- internal and external lighting and power
- hoists
- electronic measuring and testing equipment
- inspection lamps

Other equipment may also be used and this will depend on the nature and size of plant being maintained at the depot.

It is important to remember that any special requirements as a result of a particular environment are in addition to those required in BS 7671. The requirements of BS 7671 are for electrical installations, the information in Part 6, Special Installations or Locations, Particular Requirements, supplement the requirements of the remainder of the standard, they do not replace those other requirements.

Try this

In general terms, consider water and sewerage plant sites and list

a)   the type of hazards that could be encountered

b)   the types of electrical equipment that may be in use

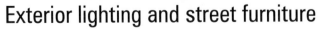

Exterior lighting and street furniture

When installing exterior lighting and equipment for use in public places there are a number of factors to be considered. Where equipment is located outdoors there is always the ravages of the weather to consider. That includes rain, hail, snow, freezing conditions, hot sunshine and solar radiation. Equipment for outside locations needs to be able to operate, under these climatic changes, without undue deterioration or reduced safety.

A further consideration is that of vandalism. Luminaires in particular are subject to abuse when placed in accessible positions. Manufacturers have developed a range of street furniture that is virtually vandal proof. An example is shown in Figure 4.18 where the bollard is made of a polyurethane coated aluminium head and steel post with clear polycarbonate beneath the louvres. The mountings are concealed with all external fasteners countersunk and tamper proof.

Figure 4.18   Lighting bollard

Another problem, particularly associated with highway street furniture, such as lamp standards, bollards and street signs, is the risk of damage as a result of collision by motor vehicles. Many manufacturers have produced designs which reduce the risks involved as a result of this kind of damage.

One example is the road bollard where the supply, control gear and lamp are contained in the base, below ground level. The light is projected up into the bollard, in much the same way as it would be from an uplighter, with a toughened lens over the lamp. In the event of a vehicle colliding with the bollard the whole of the above ground structure contains no electrical components and the base can withstand a vehicle driving over it without damage. Furthermore the base is not susceptible to the elements when exposed. Making this construction safer for everyone, the vehicle occupants, the rescue services and the general public.

Section 611 of BS 7671 deals with the additional requirements associated with Highway power supplies and Street furniture. This includes additional restrictions on methods of protection against electric shock. When undertaking installations in public areas this section should be used as reference to establish the additional requirements.

Construction sites

The construction site environment has always been accepted as being a hazardous one especially when it comes to the use of electrical equipment. For this reason reduced voltages have been recommended and used for many years.

Figure 4.19   A tower crane on a construction site will need an electrical supply

Construction sites can generally be regarded as temporary installations. There are however, some parts that are more permanent than others. Site offices, canteens, toilets blocks etc. can usually be regarded as semi-permanent so the electrical installation in these can be treated the same as any other. Where supplies are taken onto the site more care has to be taken as the risk of receiving an electric shock is greatly increased.

Section 604 of BS 7671 gives details of the additional requirements for construction sites. The maximum voltages which should be used for particular applications are given in Regulation 604-02-02. It can be seen from this information that it is not a question of a single voltage for all circumstances.

On most installations the disconnection times for an overcurrent device to operate in the event of a fault occurring is either 0.4 or 5 seconds. The minimum disconnection times for construction site installations are given in BS 7671 Table 604A. As a result of these reduced disconnection times, between 0.02 and 0.35 seconds depending on the voltage involved, lower values of Z_s are required to achieve disconnection. It is therefore quite common to find other protective measures, RCDs in particular, used on construction sites. For further information on the specific requirements for construction sites reference should be made to BS 7671 Section 604 and HS (g)141 Electrical Safety on Construction Sites, published by the Health and Safety Executive.

Try this

A mobile lighting tower is to be used to illuminate excavation work on a construction site. The tower has three 1 kW tungsten halogen luminaires mounted at 4 m. A twin, PVC steel wire armoured and sheathed cable is to be run from a switchfuse containing BS 1361 fuses at the supply intake position and will terminate at a position near the base of the floodlight. The 25 m of cable will be laid on the surface in a safe position.

a)   What is the maximum disconnection time the protection device should operate under earth fault conditions?

b)   What is the maximum value of earth fault loop impedance permissible to ensure the correct operation of a 30 mA residual current device protecting the tower supply?

Swimming pools

One area which may be considered to offer considerable increased risk to the users of the installation is a swimming pool. The quantity of water and the fact that the users are unlikely to have any additional insulation makes the risk of electric shock quite high. BS 7671 considers the requirements for swimming pool installations in Section 602. The special requirements have been classified by dividing the areas within a swimming pool into zones, very much as was the case for hazardous areas. Electrical equipment suitable for use in each zone is then identified and the location of equipment is therefore controlled.

There are several aspects to take into account, these include the:

• zone
• acceptable voltage
• degree of protection of enclosures

The zones defined around the pool determine the method of protection against electric shock. In zones A and B the only acceptable measure is SELV at a nominal voltage not exceeding 12 V a.c., rms or 30 V d.c. and generally the safety source must be located outside zones A, B and C. In zone A equipment must have a degree of protection to IPX8, in other words it can be immersed in water. In zone B the degree of protection is IPX4 unless water jets are likely to be used for cleaning purposes where this should be increased to IPX5. In both of these zones all fixed equipment must be specifically designed for use in swimming pools. Equipment in zone C (indoor pools) must have a degree of protection to at least IPX2, outdoor pools to IPX4 and where water jets are used for cleaning purposes this must be increased to IPX5. Socket outlets are allowed in zone C providing several requirements are complied with, these include:

- the use of sockets to BS 4343
- the use of residual current devices
- SELV

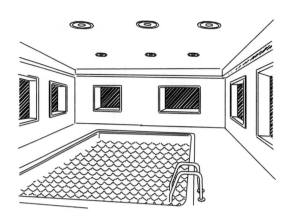

Figure 4.22   Typical small pool layout

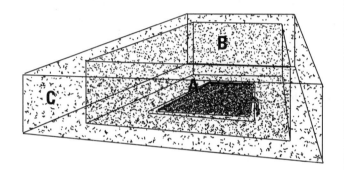

Figure 4.20   The zones defined around the pool

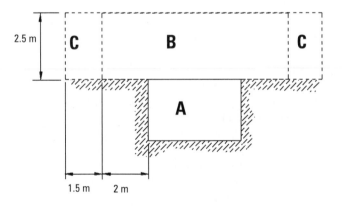

Figure 4.21   Section through zones

All exposed conductive parts in all three zones should be bonded together with protective conductors and exposed conductive parts.

As SELV circuits by definition must be earth free, the bonding does not apply to these circuits. Where there is a metal grid in a solid floor it should be connected to the local supplementary bonding.

Hot air saunas

The high temperatures and humidity within a hot air sauna make it a situation that requires special consideration. The requirements for Hot Air Saunas are covered in BS 7671 Section 603 which, like the swimming pool section, defines the particular requirements for electrical installations in these locations. The areas within the sauna have again been divided into zones. Zone A is, again, the most onerous and the only electrical equipment allowed in this area is that associated with the sauna heater. Electrical equipment in the other areas is related to the ambient temperature expected and determined by the zone dimensions. Generally the only electrical equipment inside the sauna are the sauna itself and a luminaire. The other equipment, cabling and containment system being located outside of the sauna.

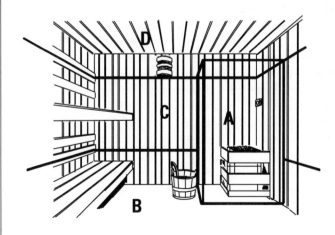

Figure 4.23   Zones around the sauna

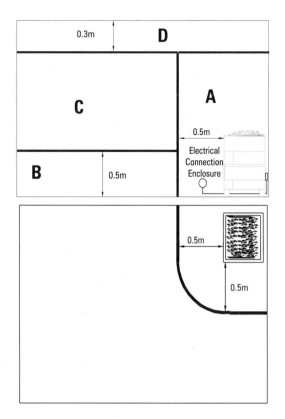

Figure 4.24   Side elevation and plan showing zones and dimensions

---

Try this

A hot air sauna is to be installed inside an existing room. There is a 100 W luminaire to be fitted inside the sauna at 2.5 m from floor level.

What are the requirements with regard to the switching of this luminaire?

Extremes of temperature

Special consideration must be given when installing cables in extreme conditions, that is extremes of heat and cold. The conditions contained within a forge area will be very different from those in a cold store or refrigeration plant. There will be a need to consider the suitability of equipment for service in these conditions and additional consideration must be given to the type of cables used. We shall be considering cables for installation under these conditions later in this chapter.

Installations involving equipment with high earth leakage current

Installations involving equipment which has a high earth leakage current, as part of the normal operation, most commonly involve data processing equipment. There are other installations such as welding, radio and radar equipment but by far the most common occurrence involves computer installations. Section 607 of BS 7671 deals with such installations and identifies the considerations which need to be made. If a circuit supplies a single item of equipment which has a high earth leakage or a number of socket outlets the considerations and requirements are included in Section 607.

Where a circuit supplies a single item of equipment then it is generally possible to establish from the manufacturer what the leakage current is likely to be. Where a number of items are supplied on a single circuit, such as a number of PCs in an office, then it may be more complex to establish the actual leakage involved. Some manufacturers may vary the phase shift on their PCs so that when there are a number connected on a single circuit the effect reduces the overall leakage current. In some cases the actual use and type of equipment may not be known, when a commercial building is constructed for lease for example, and so such information is not available to the designer. In general terms it is reasonable to expect that most commercial properties will have computers in use and it would be as well to make an allowance for the requirements included in Section 607 at the outset. In the case of an existing installation, the leakage current present could be measured with all the equipment in operation and no other circuits in use. Alternatively confirmation could be obtained from the equipment manufacturer as to the leakage current likely to occur.

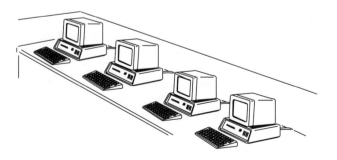

Figure 4.25

Caravans, motor caravans and caravan parks

Whilst in transit the road lighting of a touring caravan is supplied by the vehicle electrical system at 12 V d.c., as may be some of the equipment on board such as the fridge. When parked there may be some 12 V d.c. internal lighting and power, but part of the installation relies upon connection to a 230 V tap off point provided on a caravan site. This allows standard household appliances to be used such as toasters, kettles and televisions. Equipment such as the refrigerator and water heaters are often designed to work off 12 V d.c. and 230 V a.c., with some even having provision for gas supplies as well.

Figure 4.26

There are special requirements for the electrical installation within the caravan itself and further requirements for the supply on the caravan site. The details of these requirements are contained in Section 608 of BS 7671 and this Section is divided into two divisions. Division one deals with the requirements for the caravan where there are special requirements due to the construction of the caravan. A further consideration is the fact that the whole installation is mobile and subject to vibration and flexing during the towing process. Division two deals with the requirements for the site where there must be suitable provision for connection and protection against indirect contact. There are particular requirements for the protection of the socket outlets provided for connection, including the maximum grouping and the need to ensure that these sockets are protected by a suitably rated RCD. The socket outlet earth terminals must be connected to an earth electrode and should not be bonded to a pme terminal. Further information can be obtained from Section 608 of BS 7671.

The final type of installation having special requirements which we shall be considering is the type used in agricultural and horticultural locations. It is important to remember that the special considerations apply to the areas of the electrical installation which are used for the commercial activities of the business. So the farmhouse dwelling itself can be regarded as a domestic installation, whilst the farm buildings and the like would require special consideration. There could be some areas where the two may overlap depending on the nature of the site and the particular usage, the most common being the location of the earth electrode for a farmhouse supplied on a TT system.

Agricultural and horticultural installations

With regard to the use of electricity the areas in and around agricultural and horticultural installations may be divided into four main groups.

- external – no livestock present
- external – livestock present
- internal – no livestock present
- internal – livestock present

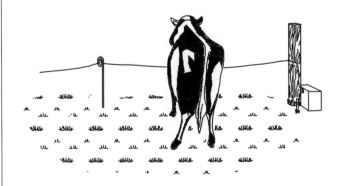

Figure 4.27

The hazards to which the electrical equipment on agricultural and horticultural installations can be exposed are considerable.

External Installations

On external installations these can include:

- weather conditions
- chemical attack
- mechanical damage
- corrosion and erosion
- use by electrically unskilled persons
- livestock abuse
- rodent and vermin attack

and we shall consider some of the most common of hazards and their typical locations.

Electricity supply between buildings

Due to the layout and requirements of this type of location the electrical supply often needs to be distributed to different buildings around the site. Installing the cables overhead is often the easiest and most cost effective method of achieving this, however the cables must be at height to prevent damage by vehicles and equipment. IEE Guidance Note 1 to BS 7671 suggest minimum installation heights for cables dependant upon the type of cable and the means of support which recommends minimum heights of 5.8 m at road crossings and 5.2 m wherever else vehicle access is possible. It is worth noting that the Guidance Notes minimum recommendations for height in locations where vehicle access is not possible are not applicable to agricultural premises. In these locations particular consideration must be given to the type of equipment and access by livestock. A suitable means of support must be used for the cables between buildings and maximum recommended span details are contained in IEE Guidance Note 1.

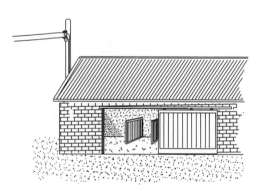

Figure 4.28   Cables between buildings

The routing of cables underground is an alternative, but these too are subject to special requirements, to reduce the risk of mechanical damage, and are more costly to install. The cable must be of a suitable construction for installation underground, with a metallic sheath or armour or it may be a concentric type. There is no recommended depth for buried cables given in BS 7671 or the guidance material which accompanies it. As a rule of thumb guide 0.5 m is considered

suitable for general purposes and 1.0m if the soil is likely to be cultivated. These are however only general guides and consideration should be given to the use of the land and the risk of damage. Cables should be covered by cable tiles or marker tape to provide warning to those excavating in the area.

Electrical equipment

The general environment means that equipment should also be constructed to withstand external influences such as:

- dust
- water ingress
- corrosion
- mechanical stress
- flora
- fauna
- vermin attack
- solar radiation
- wind

To keep the risks from electrical equipment to a minimum Section 605 of BS 7671 gives precautions that should be adopted. For example, electrical equipment which is for normal use should have a minimum IP rating of at least IP44. Generally all circuits supplying socket-outlets must be protected by a residual current protection device. Protection against indirect contact due to the adverse conditions must be designed so that the maximum disconnection time under fault conditions does not exceed 0.2 seconds. To ensure this requirement can be met new tables for Z_s are given in Tables 605B1 and B2.

Figure 4.29   Milking parlour

The nature of the work means that it is often necessary to have outside illumination at several different points. The luminaires must always be suitable for the environment they are being used in and positioned so that they are not exposed to mechanical damage.

Remember
Supplies to equipment via flexible cables can be a particular hazard and special care should be taken when such connections are undertaken.

Electric fences

An electric fence controller comprises a base unit which pulses a high voltage between its output and earth. The output is connected to a bare conductor which is run as a fence wire, normally 600 mm above the ground on insulated stakes around the area to be controlled. The control unit is normally mounted above ground level at one end of the run and the connection to earth is via an earth stake. The controller may be either mains or battery operated. The battery operated type are often used where there is no mains supply readily available at the desired location. When livestock come into contact with the bare conductor they receive the pulse shock as they are in constant contact with earth and hence complete the circuit.

Mains operated fence controllers should comply with the requirements of BS EN 61011, and battery operated units which may be connected to a mains supply should comply with BS EN 61011-1. The purpose of these constraints are to ensure that the high voltage output of the unit cannot come into contact with any other electrical supply or equipment. The reasons for this are that the discharge voltages on electric fences can be as high as 10 kV to earth.

Figure 4.30   Electric fence

Internal installations

The general environment of the agricultural and horticultural installations means that many of the factors that apply to outside installations also apply to internal ones. Disconnection times and Z_s values along with the installation factors are an example.

Equipment should be constructed to withstand environmental influences such as:

- dust
- dampness
- corrosion
- mechanical stress
- flora and fauna
- vermin attack

Wherever possible electrical equipment should be placed so as to be inaccessible to livestock either by location or protection. The selection of suitable equipment is essential and it should always be located, as far as is practicable, to minimise the detrimental effects of the environment.

In some types of premises as the temperature drops with the cold weather it is necessary to have heating on pipes and some containers. The type of heating used for this is generally low power tape heaters of about 4 Watts/metre, that are wrapped round the surface to be kept warm. These are usually supplied from double wound transformers and controlled by frost stats. As the temperature drops below a predetermined level the supply is switched on to the heaters. As the pipes are usually well lagged there should be clear labels stating that the pipe is protected by electrically supplied heaters.

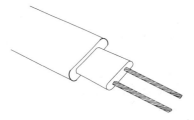

Figure 4.31   Low power tape heater

Floor and soil warming

Farming installations occasionally require floor warming systems to be installed. Horticultural installations may require soil warming for plant propagation. Both types of installation have to be installed to similar standards. The heating cable should be one designed for the purpose, having protection from dampness and corrosion. They must be installed to manufacturers' instructions ensuring that they are not exposed to any mechanical damage either during or after installation. They must be also be completely embedded in the substance they are designed to heat. Where there is likely to be any movement in the soil then precautions have to be taken to ensure the heating cable cannot be damaged. Designated maximum conductor operating temperatures are laid down for different types of floor warming cables. These are shown in Table 55C of BS 7671.

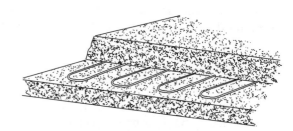

Figure 4.32   Typical soil warming system

Try this

What is the maximum conductor operating temperature for a floor-warming cable of

a)   heat-resisting PVC over conductor

b)   general purpose PVC over conductor

c)   silicone-treated woven-glass sleeve over conductor

69

Earthing and bonding

Livestock is very susceptible to electric shock so wherever they are present the equipotential bonding requirements are very strict. Generally, where livestock is kept, supplementary equipotential bonding is required to connect all exposed conductive and extraneous conductive parts together. This includes metal grids that have been installed in floor areas for supplementary bonding. There is also a reduction in resistance of the bonding conductor as defined in BS 7671, Regulation 605-08-01.

Where residual current devices are used for protection against fire their operating current should not exceed 500 mA.

Having considered some of the special locations which we may encounter, we should look at the types of cable available for adverse conditions. This will enable a suitable selection to be made which is appropriate to the particular conditions, whether it be due to the environment or to particular performance requirements.

Try this

A building or shed used for rearing livestock is to have a number of lighting outlets and one socket outlet installed. A consumer control unit is to be fitted in the barn for the distribution. The supply is to be by overhead cables from another building.

a) Assuming that the building is to be protected by a 30 mA residual current device, what is the maximum permissible earth electrode resistance?

b) What consideration should be given to the position of the earth electrode associated with this installation?

c) Considering that the main purpose of this building is for housing livestock, what considerations, particular to agricultural and horticultural premises, should be given to the electrical installation?

d) What special requirements apply to the installation of the socket outlet in this building?

Cables for special installations

Each installation will have a set of criteria for which the cables and equipment must be suitable. This criteria may be due to the

- environment in which the cables and equipment are to be installed
- conditions which are or may be present
- performance criteria for use under adverse conditions

The selection of an unsuitable cable or type of equipment can introduce a further hazard to the installation, the users and building fabric.

Environmental and performance requirements may affect the particular properties for a cable in a number of different ways. For example, very low temperatures can cause insulation materials to become brittle and crack or break. High temperatures can cause the insulation to become soft and melt and possibly give off dangerous emissions. The environment in which cables are installed and required to operate can also be inhospitable to the cable. We may need to use cables that are able to resist water, acid, alkalis and other chemicals or that are able to continue to operate in fire conditions, all without causing further hazards.

Whilst we have considered some of the installations which require special consideration there are some others which introduce their own particular hazards which will affect the choice of cable for the installation. These include;

cold:
cold stores, refrigeration plant, and exposed locations

heat:
boiler houses, bakeries, steel works and kilns

damp:
laundries, car washes, marinas and exposed locations

dust:
wood machining, flour mills, cement works and material production

corrosion:
diesel storage and dispensing, oil stores, electroplating and pine stripping vats

static:
medical areas (operating theatres), electronic equipment production and gas operations

All of these, and others not listed, produce particular hazards for the installation of electrical equipment and cables. In view of this we shall consider the properties of some of the cables available and so determine their suitability for particular installation requirements.

The construction of a cable will be dependant upon its intended use and method of installation. The cable may consist of one or all of the following; conductor, insulation, sheath, mechanical protection and over-sheath. It may be that the mechanical protection is in the form of a cable sheath made from an insulating material.

Figure 4.33

When a cable is to be installed in a hazardous environment, then the whole of the cable must be suitable for the conditions which exist. For example it would be inappropriate to install a cable in an area where the ambient temperature is say, 90 °C, if the insulation is only suitable for 70 °C, whilst the outer sheath is designed for a maximum of 120 °C. We must therefore give consideration to the construction of the cable including the individual components and the cable as a whole.

Where cables are to be installed in areas where organic solvents are present then the outer sheath of the cable will need to be able to withstand the solvents. As the conductor insulation should not come into contact with the solvent, the insulation may not need to be able to withstand it.

There are many materials which are used for insulation and sheathing, some of which we shall consider here. These materials can be roughly divided into two main groups, the thermoplastic and the thermosetting types.

Thermoplastics

Polyvinyl chloride (PVC)
These are thermoplastics and are a class of materials rather than a single product. The exact properties of the material depend on the quantities of plasticizer and filler in the compound, both of which are adjustable over a wide range.

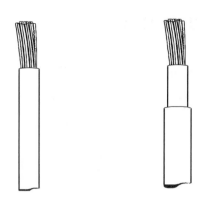

Figure 4.34   PVC cables

General purpose PVC when used as an insulator has a maximum conductor operating temperature of 70 °C. However cables are available which will allow the conductor to operate at temperatures up to 90 °C. As a rule PVC cables should not be handled or installed at temperatures below 0 °C, although some compounds are suitable for temperatures as low as −10 °C.

PVC offers good resistance to water and salt water solutions, acids and alkalis. It only offers moderate resistance, if any, to organic solvents and oils. Special additives can be used to give the compound resistance against chemical solutions, if a demand is identified.

PVC is not only used as an insulator but also used as the outer sheath on many types of cable. Apart from the temperature restrictions there are also problems should the general material be involved in a fire. The fumes that are given off are highly toxic and contain hydrogen chloride which, when combined with moisture, forms hydrochloric acid. To overcome this compounds have been developed that use cross linked processes to create materials with different properties.

Cross-linked polyvinylidene fluoride is an example of a compound designed to give a low smoke emission and virtually no halogen gases. The resistance to water and acids are very much the same as general PVC compound, but extra care may need to be taken when bending and the minimum radii may need to be slightly greater.

Polyethylene (PE)
This is the other main group of thermoplastics and is more commonly known as polythene. It is produced from the gas ethylene, a by-product of the petroleum industry, which is heated to about 150 °C under tremendous pressure. There are two types used for electrical cables, Low density (LDPE) and High density (HDPE). Both of these have good electrical properties as well as a low water uptake. They also have excellent resistance to water, inorganic salts, acids and alkalis, as well as good resistance to organic solvents. The high density (HDPE) is less flexible than the lower density type. The conductor operating temperature of LDPE is 70 °C where HDPE is 80 °C.

Thermosetting plastics

Cross-linked polyethylene (XLPE)

This is a thermosetting material that has a molecular structure which is not simply linear. Unlike thermoplastics, it has a large number of cross links making up a multi-directional chain. This creates a more rigid structure with greatly increased resistance to deformation of insulants at higher temperatures. This gives a conductor operating temperature of 90 °C, which is higher than high pressure polyethylene. Its insulation resistance, being higher than PVC, enables its thickness to be reduced. It also offers excellent resistance against water, inorganic salts, acids, alkalis, and good resistance to organic solvents.

Figure 4.35   XLPE

Ethylene Propylene Rubber (EPR)

This is a very complex thermosetting insulating material formed by the combination of a large number of components. EPR has excellent anti-ageing properties and offers good resistance to weather, ozone, acids and alkalis, but is less resistant to oils and solvents. Other terminology which may be encountered includes EPDM for terpolymers of ethylene, propylene and a small proportion of diene; and EPM for ethylene propylene copolymers.

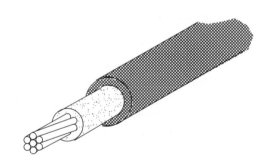

Figure 4.36   Heavy duty ethylene propylene rubber (EPR)

The maximum conductor operating temperature is 90 °C.

Other poly type insulations include:

Polypropylene (PP)

This has a higher stress cracking resistance than PE but is less flexible. It offers excellent resistance against water, inorganic salts, acids and alkalis, it also has good resistance to organic solvents.

Polyfluoroethylene-propylene (FEP)

In general this offers good ozone and weather resistance, together with excellent chemical resistance.

Polyvinyldene-fluoride (PVDF)

This also offers good ozone and weather resistance, and protection against most chemicals.

Polytetrafluoroethylene (PTFE)

This, like many of the plastics, is better known by its letters than its name. This is created from a gas prepared from the raw materials fluorspar and chloroform. Processing of the material is difficult, as before it can be extruded it has to first be sintered. This is a heating and pressing technique that fuses together a mass of tiny particles. It can then be extruded as an insulating sleeve over electrical conductors. The high temperatures used in the extrusion processing means the copper conductors must be nickel or silver plated to stop severe oxidisation. The chemical and thermal properties of this are excellent giving a temperature range of between –70 °C and 250 °C.

Figure 4.37   PTFE

Polychlorotrifluoroethylene (PCTFE)

This is similar to PTFE but is easier for the manufacturers to work with. It offers good ozone and weather resistance but may swell in some organic solvents.

Ethylene-terafluorethylenecopolymer (ETFE)

This is also used as a substitute for PTFE offering similar properties but is cheaper to produce. It will not, however, stand up to the very high temperatures of PTFE.

Ethylene-chlorotrifluoroethylene copolymer (ECTFE)

Like the other fluoroethylene compounds this also offers good resistance to ozone, water and most chemicals.

Polymide (PI)

Although this is a "poly" type material, it has poor weather resistance and only moderate resistance to most chemicals.

Polyurethane (PUR)

As cable insulation this offers excellent resistance to abrasion at the same time giving good flexibility and oil protection. It only offers moderate resistance to organic solvents and very poor resistance to acid and alkalis.

Polystyrene (PS)

The electrical properties of this are good, as is the resistance to water, acids, alkalis and salts. There is no resistance to organic solvents.

In addition to all the poly-based materials there are several that are rubber based.

Natural rubber (NR)

This has excellent tear resistance but it does not stand up to the weather and ozone conditions. Unlike many of the poly materials it remains flexible in cold temperatures. The resistance to water, acids, alkalis and salts is good but it has little or no resistance to certain solvents.

Silicon rubber (SiR)

Silicon rubber is unique, being the only synthetic elastomer with an inorganic chemical structure based on silicon and oxygen. Silicon insulated and sheathed cables are capable of working at extremes of temperature as far apart as –60 °C up to 150 °C. At low temperatures they retain their flexibility and in excessive heat, such as in a fire, the silicon changes to form an insulating layer which, in the short term, will maintain the cable's integrity – important for lighting and fire alarm systems. This is more expensive than other elastomers, consequently it is used on special applications involving high or low temperatures.

Styrene-butadiene rubber (SBR)

Like natural rubber this remains flexible in cold conditions, but has poor ozone and weather resistance. The resistance to water, acids, alkalis and salts is good but again there is no resistance to certain solvents.

Butyl rubber (IIR)

This has good anti-ageing properties and remains flexible in cold conditions. It also offers good resistance to inorganic salt water, acids and alkalis.

Polychloroprene-rubber (CR)

This offers good ozone and weather protection together with very good resistance to acids, alkalis and salts.

Chlorosulphonated polyethylene (CSP)

Despite the poor tear resistance of this material it is excellent for standing up to ozone. It also offers good resistance to inorganic salt water solutions, acids and alkalis.

Ethylene-vinylacetatecopolymer (EVA)

In cold weather the flexibility of this is poor but it offers very good resistance to ozone and the weather. Although its resistance to acids and alkalis is good this is reduced when it is exposed to oils and solvents.

Alternative insulators

Cables can also be insulated with materials such as magnesium oxide and glass.

Mineral insulated (MI)

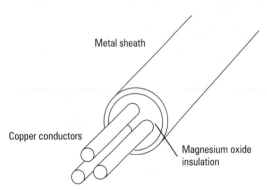

Figure 4.38   Mineral insulated metal sheathed cable

Conductors insulated with magnesium oxide are sheathed in metal to create a compact and strong cable. If magnesium oxide is exposed to the atmosphere it will absorb moisture and its insulating properties are impaired. For this reason a special seal must be used at the end of each piece of cable. When the cable is properly terminated it forms a complete metallic enclosure which offers great mechanical strength, water resistance, is fire proof and capable of operating at high temperatures up to 250 °C.

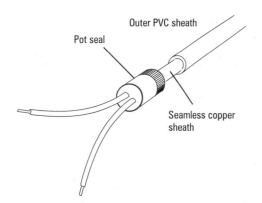

Figure 4.39   PVC sheathed MICC cable

Often an extruded covering of PVC compound is used to protect the metallic sheath, but without this the cable offers no halogen gas or smoke emission. When cables are used at high temperatures the seal used at the cable ends must be capable of withstanding the maximum operating or ambient temperature.

Glass

A glass fibre braid is sometimes used over silicone rubber to provide added protection. This has a working temperature rating of 150 °C with a short term of up to 180 °C.

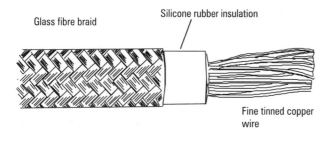

Figure 4.40

There are many combinations of insulation and sheathing to cater for different applications. When there are special situations the exact make up of the cable must satisfy the worst conditions it is likely to encounter.

Corrosion and erosion

There are many synthetic materials used in electrical installation work, however conductors and many enclosures are still made of metal. Such enclosures and protective conductors are often made of steel, such as conduit, trunking and armour, and so may be subject to corrosion. Conductors do not only include the phases and neutrals but also the protective circuits. Whereas the live conductors are made of copper or aluminium, protective conductors often include steel.

Where an electrical installation is in a dry atmosphere without being subjected to water spray the problems of corrosion are minimal. However if the installation is carried out in a hostile environment, say a cow shed, then the risk of corrosion is very high. The combination of damp and acidity is highly corrosive and in these conditions steel will rust rapidly and copper will oxidise and may even sulphate, if not suitably protected.

Figure 4.41   Steel conduit and trunking may corrode if placed in a damp acidic atmosphere without protection

Corrosion and erosion can be considered from two different points of view. Metals can be corroded or eroded by chemicals "burning" into the surface and hence reducing the thickness of the material. The other source of corrosion is due to the electrolytic action between different materials. The galvanic series shown in Table 4.6 lists a selection of materials that react to each other. Basically the further they are apart in the series, the more electrolytic action takes place between them.

Table 4.6 Galvanic series

Anodic	Least Noble
(Corroded end)	
magnesium	
zinc	
aluminium alloys with magnesium	
pure aluminium	
cadmium	
aluminium alloys with copper - Duralumin	
mild steel	
cast iron	
lead	
tin	
nickel	
brasses	
copper	
bronzes	
silver	
graphite	
gold	
Cathodic	**Most Noble**

An example of the result of the galvanic series is the dry Leclanché cell where zinc and graphite are the active components. These materials are at opposite ends of the galvanic series and the result is an e.m.f. of 1.5 V. The electron flow is from the zinc to the graphite, leaving the zinc very much reduced in thickness.

Another example of this can often be seen where copper pipes are connected to steel radiators. The steel of the radiator gives itself up through the water, which acts as an electrolyte, to the copper. The result is that eventually the radiator corrodes on the inside and shows signs of leaking around the connection. Copper and steel are not as far apart in the galvanic series as graphite and zinc, so the flow between them is less and the corrosion takes longer. The rate of deterioration of the radiator also depends on the chemicals in the water, and hence the effectiveness of the electrolyte.

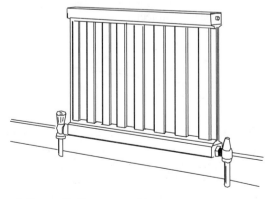

Figure 4.42   Radiator connections

Protection against corrosion and erosion

Corrosion protection is usually carried out using either a coating process, or the use of an electron reversal system.

Coating or cladding

The use of coating or cladding of metal is the most common method of protection against corrosion. The coating or cladding is used to isolate the active materials, either from each other or the effective electrolyte. A large number of metal pipes and ducts are made from steel which corrodes into rust when given a suitable environment. The corrosion of steel is due in part to the electrolytic action between different particles in the material's structure.

When carbon steel is made, microscopically thin layers of ferrite and cementite are arranged alternately in the structure. Ferrite is anodic to cementite and so given the right conditions, the ferrite corrodes away leaving the cementite layers. As these remaining layers are very brittle, they soon start to break away. As the particles cannot be isolated from each other protection must stop the introduction of an electrolyte.

There are several methods used to protect the surface of steel to stop or reduce the effect of corrosion. Many of the methods used to protect steel are also suitable to provide protection for other metals.

Painting

This is used extensively to reduce atmospheric corrosion and also provides a decorative finish. The best results are obtained if the surfaces of the steel are first phosphated. This involves treating it with a phosphoric acid preparation, which not only dissolves rust, but also coats the surface of the steel with a dense, slightly rough surface of iron phosphate. The surface is then partially protected from corrosion and is ideal to receive the paint primer. The paints used are generally polymers or epoxy powders, although aluminium primers are sometimes used when painting in situ.

> *Remember*
> **Corrosion** – the wearing away of metals especially by chemical action, for example rusting.
>
> **Erosion** – wearing away by the action of water, wind and physical contact with moving objects.

Metallic coating

One metal can be used to protect another from corrosion. In this process a thin coating of a corrosion resistant metal is applied to the surface of the one to be protected. The method of coating and the metal used can vary depending on the application and environment to be encountered.

- Hot dipping is used to coat iron or steel with either tin or zinc. This process involves passing clean sheets of the metal through a bath of molten tin or zinc and then through rollers to remove the surplus material. When zinc is used on mild steel, this is usually referred to as galvanising.
- Spraying can be used to coat surfaces with a wide range of molten metals. Zinc is the most commonly used.
- Sherardising is a cementation process. The steel components to be protected are rotated and heated in a drum containing zinc powder at about 370 °C. This deposits a thin uniform layer of zinc on the exposed surface of all of the components.
- Electroplating can be used to deposit a large variety of metals on different surfaces. The electrolytes and currents required vary depending on the materials used and the surfaces to be plated.

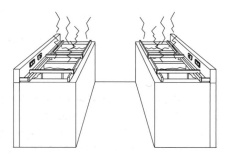

Figure 4.43   Electro plating baths can give off corrosive fumes

- Cladding is used on sheet metals and cables. On sheet metals the basic metal is sandwiched between sheets of the coating metal. The sandwich is then rolled to the required thickness and in the process the coating film welds onto the base metal. In a similar way aluminium cables can be clad with a thin film of copper to add strength as well as reduce the corrosive effects.

Protection by oxide coatings

Some metals naturally form a dense film of oxide on their surface. This then creates a protective barrier so the metal is not corroded further. Both copper and aluminium possess these properties to some extent. Examples of copper oxidisation can be seen on roofs of buildings where copper sheet has been used. In these cases instead of the usual copper colour, the roofs take on a light green colour. If this was to be cleaned off a fresh surface of oxide film would form. If this was continued, the metal would gradually be oxidised and worn away.

Aluminium also naturally forms an oxide film to protect itself. In this case there is a process known as anodising that can be used to make the oxide film thicker and offer greater protection. To do this the aluminium is placed in an electrolytic bath and connected as an anode. As current is passed through the bath atoms of oxygen are liberated at the surface of the metal, these immediately combine with the aluminium creating a thicker oxide film.

76

Cathodic protection

It is sometimes necessary to fit metal pipes and structures in positions where electrons will flow naturally from the steel to the surroundings. This is common in buried pipes where the steel will give itself up to the soil and corrode. If the flow of electrons could be reversed then the loss of the steel to the soil would be reduced. This situation can be created by using electrodes of magnesium or zinc buried adjacent to the pipe. As these materials are more anodic than steel, they deteriorate instead of the pipe they protect. For this reason they are called sacrificial anodes.

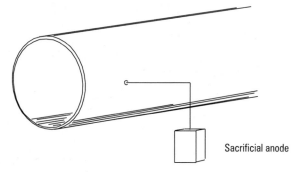

Sacrificial anode

Figure 4.44   Pipe and sacrificial anode

Similar results can be obtained using a small d.c. supply connected between the pipe and earth electrodes. This is basically a current set up so that a flow of electrons passes from the electrode through the earth and to the protected steel. The voltage of the d.c. supply has only to be high enough to overcome that naturally found in the soil plus a little more. The natural voltages are usually somewhere between 0.4 V and 0.75 V, depending on the make-up of the soil.

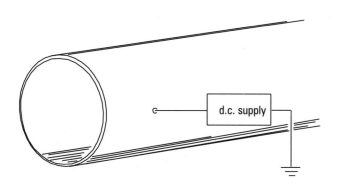

Figure 4.45   Pipe, d.c. supply and electrode

Figure 4.45 shows the basic circuit used but in practice one supply unit would serve a number of connections between the pipe and separate electrodes. This method of protection is known as an impressed current system. The d.c. supply used for this may be from a battery, transformer rectifier unit or even solar cells depending on circumstances. A typical circuit arrangement is shown in Figure 4.46.

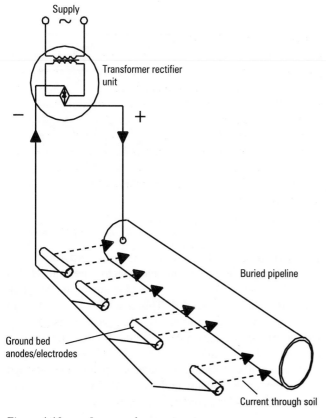

Figure 4.46   Impressed current system

To determine the actual voltage that is naturally existing, and to check after protection has been connected, a special test must be carried out. This involves the use of a very high resistance voltmeter and a device known as a half-cell. These may be built into one unit for practical test purposes or two separate units as shown in Figure 4.47.

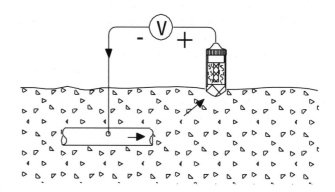

Figure 4.47   Diagram showing the circuit test using a half-cell

The half-cell is made so that one end can be placed onto the soil close to the buried metal. When it is placed in this way, the voltage can be measured directly on the high resistance voltmeter. The amount of deflection on the meter indicates the state of the electron flow between the soil and the pipe.

Half-cells vary in detail depending on their exact application. Figure 4.48 shows a typical one suitable for taking the measurements as described above.

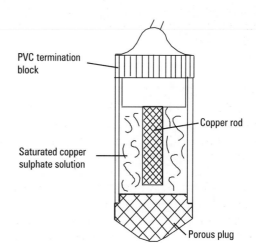

Figure 4.48   Half-cell

Fire alarm systems

Having considered hazardous areas and installations which require special consideration we shall now look at systems designed to protect the user and property from other risks. We shall begin by considering fire alarm and detection systems. The likelihood of fire can never be ruled out but, with good planning and a reliable fire detection and alarm system, any resulting losses can be kept to a minimum.

One essential feature of a fire detection system is its speed of response to the symptoms of fire or a situation which can result in fire.

Human response is quick and sensitive to a wide range of pre-fire indicators, such as heat, smell and visual effects. A person noticing any of these indications can respond immediately, however they must have ready access to some form of signalling device. The means of signalling must allow the information to be transmitted quickly and effectively in order that the appropriate action can be taken promptly. Constant human presence is not normally a practical solution for fire protection. Premises are unlikely to be supervised twenty four hours a day, seven days a week but the risk of fire is ever-present.

Where human supervision cannot be guaranteed, automatic fire detection equipment will need to be installed if fire protection is to be provided. The effectiveness of the protection is a matter of good design and the correct choice of equipment.

BS 5839 provides recommendations for the planning, design and installation of fire detection and alarm equipment. It classifies the type of system, makes recommendations for the installation and commissioning of the equipment and lists the responsibilities of the user.

Figure 4.49   Break glass alarm contact

Types of detection and alarm system are categorised dependant upon what the system is designed to protect and the extent of the cover provided. We shall consider the classification of the systems and then look at the wiring systems employed.

Types of system

These are classified under the headings:

P	Protection of property
L	Protection of life

and

M	Manual systems

These are further sub-divided into:

P1 and L1	Systems installed throughout the protected building.
P2 and L2	Systems installed only in specified parts of the protected building. Normally covers L3 system.
L3	Systems installed only for the protection of escape routes.
M	Manual systems have no sub-division.

Types of circuit

Open circuit system

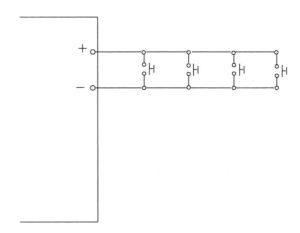

Figure 4.50   Open circuit system

This is a two-wire system in which the alarm is activated when connection is made. Call points and detectors are connected in parallel and when the circuit is closed at any of the detection points the alarm is sounded. This is a very simple system which incorporates a minimum of components. It is however vulnerable to defects which could render the system inoperable. For example, a broken cable or loose connection could leave all or some of the detection points disconnected and a fire could remain undiscovered. Similarly, a dirty contact at any of these points would fail to close the circuit and the call would remain undetected.

Monitored open circuit system

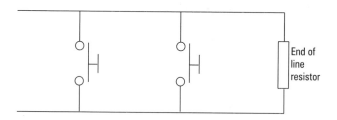

Figure 4.51   Monitored open circuit system

To improve the reliability of the open circuit system the detector circuit is terminated in an end of line resistor. This ensures that a small monitoring current flows through the circuit at all times so long as the condition is healthy. If the circuit is broken at any point, the monitoring current will cease to flow and a detecting device will respond by indicating a fault condition on the control panel. This will not be a full alarm but a supervisory buzzer and signal lamp will attract attention to the condition.

Closed circuit system

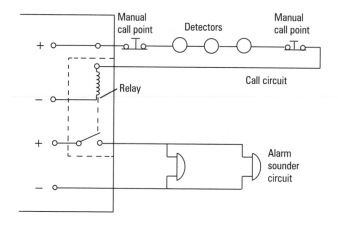

Figure 4.52   Closed circuit system

In the closed circuit system the circuit opens when an alarm is initiated. A break in the circuit operates a relay which in turn sounds the alarm. This system is to some extent "self monitoring" as a broken conductor will cause the alarm to sound. The call circuit can be extended to incorporate a number of detecting devices so long as this is consistent with good design. If however a short circuit occurs between the conductors of the detecting circuit then a possible fire alarm indication could remain undetected.

Monitored closed circuit

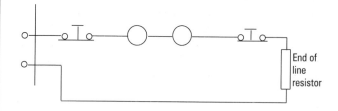

Figure 4.53   Monitored closed circuit system

An "end of line" resistor in the detection system will ensure that a low monitoring current flows continuously throughout the loop. A break in the circuit will cause this current to cease and an alarm will be initiated. A short circuit which would otherwise result in a failure of the system would produce a noticeable increase in the monitoring current and a system failure would be indicated at the control panel.

Try this

In a building that you know well, without disturbing anything, check out the fire alarm system installed. List below all the equipment that you can see.

Principles of fire detection

Having considered the types of circuits used we need to look at the types of devices used for fire detection. We shall begin with the simple break glass unit. This is not strictly a detector, as it provides a means for people to trigger the alarm system, and performs no automatic sensory function.

Manual detectors

The commonest form of this type of detector is the manual break glass device in which the contacts are held against a spring by means of a glass pane mounted in front of a button. These devices can be normally open or normally closed depending on the system employed.

Figure 4.54   Break glass alarm contact

To operate the alarm the glass is broken, either by hand pressure or by means of a small hammer attached to the device. When the glass is broken, the contacts assume the alarm position. It is often the case that a person, on discovering a fire, will break the glass and then press the button inside. This could replace the contacts in the safe position. The design of the system should be such that this action does not cancel the alarm.

Manual alarm lever

The pull-down station incorporates a breakable perspex rod instead of a window and once the device has been operated it cannot be reinstated until it has been dismantled, and the rod replaced.

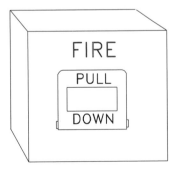

Figure 4.55   Manual alarm lever

Automatic detection devices

The detectors being considered here are examples of the types most commonly used. There are others which may be used for special applications or specified for other reasons. It would be a good idea to consult manufacturer's literature for further examples and applications.

1) Smoke detectors

There are two main types of smoke detector.

Figure 4.56   Ionisation smoke detector

One group is the ionisation type. These contains two chambers, each chamber containing ionised air surrounding positive and negative electrodes. A small pellet of radioactive material in each chamber is used to ionize the air. One chamber is sealed and this is used as a standard. The other is open to the atmosphere. A small current will flow in both chambers as the positive and negative ions are attracted to the electrodes of the opposite polarity. As long as the air in the open chamber remains clean, the current flow will be approximately equal to that in the sealed chamber.

If smoke finds its way into the open chamber its presence will interrupt the flow of ions and the current is reduced. A sensitive comparator device is triggered by this current imbalance and an alarm signal is initiated.

Low-cost, fully independent ionisation smoke detectors are to be found in many homes. These may be powered by a small dry battery and have a self contained alarm sounder. When these are suitably located they are loud enough to wake most people from a deep sleep.

The other main group of smoke detectors are of the optical type which employ a light source and a photo-electric cell.

Figure 4.57   Photo-cell smoke detector

It would be natural to assume that the presence of smoke would obscure the light from the lamp and the loss of signal from the photo-cell would result in an alarm condition. Although smoke detectors of this type have been used, it may be found that a more sensitive variety makes use of what is known as the "Tyndall Effect".

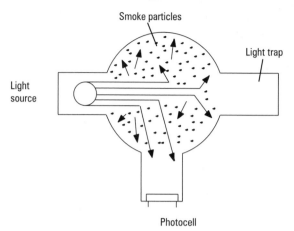

Figure 4.58   The "Tyndall" effect

The light beam is directed across the chamber into a non-reflective receptacle. A photo-electric cell is situated at right angles to the light beam and under healthy conditions no light will fall on its sensor. If particles of smoke enter the chamber, light is reflected from these which is detected by the cell and the alarm signal is transmitted by the device.

Remember

There are two main types of smoke detector:

ionisation type and

optical type

Smoke detectors should never be used where there is likely to be smoke or fumes present as part of the normal environment, for example car parks, kitchens and boiler rooms.

Ionisation smoke detectors situated in very draughty conditions may give false alarms – special precautions may need to be taken.

Temperature rise detectors

As in the case of smoke detectors, there are two basic types within this category.

The fixed temperature thermal detector, which may be either ;
- a fusible element which melts when the temperature reaches the melting point of the alloy

or
- a bi-metallic device which opens a set of contacts when a predetermined temperature is reached.

Whilst these devices are reliable they do not make any allowance for changes in ambient temperature.

Figure 4.59   Temperature rise detector

Rate of rise detection

The other main type of heat detector is the "rate of rise" type. This type normally employs two bi-metallic elements, one of which is more sensitive than the other.

If the rate of temperature rise is slow, such as the change from night time to daytime temperatures in an office, the device will register no change. This is because the elements will remain in agreement as the change is quite gradual.

If the change in temperature is rapid, such as occurs when a fire breaks out, the fast element will respond before the slow element has started to deflect. This will cause one of the pairs of contacts to operate and thus operate the alarm system. It is likely that a fixed temperature detector will be incorporated in this device. This will ensure that the device will eventually operate, at a predetermined level, irrespective as to how gradual the rise of temperature is.

Remember

There are two main types of heat detector:

fixed temperature type and

rate of rise type

Try this

List the following:
1. Types of circuits used for fire detection systems.

2. Three types of alarm initiating devices.

Table 4.7 Suitable locations for smoke detectors

No.	Location	Ionisation	Photo-electric	Reason for choice of #
1	Corridor/Walkway		#	Air current may exist
2	Stairway		#	Air current may exist
3	Elevator Shaft/Duct		#	Air current is present; smouldering is probable
4	General Office Room/Day Room		#	
	Conference Room		#	
	Waiting Room		#	
	Hotel Guest Room/Hospital Ward		#	Smouldering is probable
	Hotel Foyer		#	
5	Dining Room		#	
6	Lobby Hall	○	#	
7	Department Store/Market		#	
8	Theatre Stages and Audience Hall		#	Use of theatrical smoke may present a problem
9	Substreet/Walkway		#	Air current exists
10	Warehouse		#	Not if diesel or propane fork lift trucks are operating in area
11	School		#	
12	Library Room	○	#	
13	Public Meeting Hall/Gym		#	
14	Stage Setting Room		#	
15	Clinic Room		#	
	Nursery Room		#	
	Treatment/Operation Room/Child Care Room/Laboratory		#	Prevents false alarm due to use of naked flame
	X-Ray/Treatment Room		#	Radioactivity
16	Photo Studio/Beauty Parlour		#	
17	Dark/Developing/Copying Room		#	Presence of gaseous substances
18	Studio/Recording Room		#	Possibility of a flaming fire (optionally flame detector)
19	Machinery/Electrical Room		#	Possibility of ionised air
20	Factory/Workshop		#	
21	Church/Chapel		#	
22	Telephone Exchange Room	○	#	Smouldering fire
23	Cargo Handling Area		#	Presence of air current and dust
24	Spirit Fuel Stove	○		Fast, clean burning fire (optionally flame detector)

Key: # = most acceptable detector ○ = acceptable detector, though not necessarily the best
Reproduced with kind permission from Menvier, Division of Cooper Lighting and Security Ltd

Flame detectors

The flame detector recognises the presence of fire by detecting infra-red or ultra-violet radiation given off by the flame source. The detector responds to a "flicker frequency" between 5 and 30 Hz which is given off by flames. This avoids false alarms caused by sunlight or other legitimate sources such as heaters. The flame detector is set to scan a given area and the radiation is focused on to an infra-red sensitive photo cell. Some types of flame detectors have a fixed head which monitors a certain area within the field of vision covered by a lens. Other types have a moving head which rotates through 360 degrees and therefore monitors a much larger area.

A filter circuit automatically rejects all steady state infra-red radiation. Any radiation which falls within the "flicker frequency" band will be recognised and the alarm will be initiated. Ultra-violet flame detectors must also be designed to reject radiation from sunlight or other sources such as discharge lighting installations.

Flame detectors are unable to detect smouldering sources and should only be used in specialized applications or in conjunction with other automatic forms of detection. They do however respond more quickly than other forms of detection because they do not depend on the transportation of the products of fire from the source to the detector. They can provide surveillance over a very large area such as stores or warehouses. Whilst they need not be ceiling mounted, it is important that they have a clear line of sight over the area to be protected.

The siting of call points

The location of call and detector points plays a vital part in providing suitable and effective protection.

Manual call points should be situated on all exit routes and in particular on landings and stairwells. No person should have to travel more than 30 m in order to reach a call point. In some situations where the mobility of persons is impaired, such as old peoples' homes, the distance should be suitably reduced. Where there are particular fire hazards within a building, a call point should be placed close to the hazard.

The recommended mounting height is 1.4 m above finished floor level with easy access, good lighting (including under emergency lighting conditions) and fixed to a background of contrasting colour. They may be flush mounted if seen from the front only but if seen from the side they must have a visible profile area of at least 750 sq.mm.

Smoke and heat detectors

Smoke and heat will normally be concentrated in the highest parts of the building and this is the best position for such devices. A smoke detector is best sited within 600 mm of the ceiling and a heat detector between 25 and 150 mm. Typical layouts for open flat ceiling areas are as shown in the following illustrations.

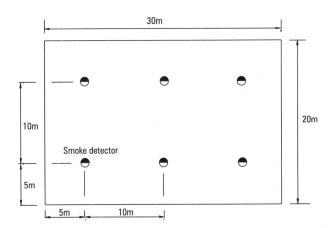

Figure 4.60   Typical layout of smoke detector system

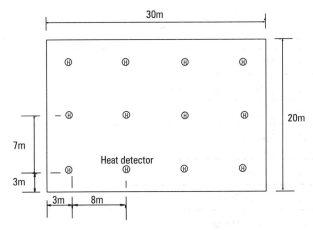

Figure 4.61   Typical layout of heat detector system

For pitched ceiling areas, smoke detectors are best sighted in the apex. Flat ceilings with drop beams require care to ensure that detectors are located in all areas where smoke is likely to be concentrated.

In general terms;
- A smoke detector will cover an area of 100 sq.metres.
- A heat detector will cover a smaller area, probably no more than 50 sq. metres.

The maximum distance from any point on the ceiling to the nearest smoke detector will be approximately 7.5 m and 5.3 m for heat detectors.

As smoke and gas rise from a fire they may become diluted by clean cool air already present in the room. This may not be a problem with low ceiling heights, but as the height increases additional care is required. Additional steps may have to be taken in such circumstances to ensure that the system is sensitive to the signs of fire.

Heat detectors may be used with ceiling heights up to 9 m. High temperature heat detectors are only reliable for quick response at heights up to 6 m. Optical smoke detectors can be installed on ceilings up to 25 m.

Where fire fighting equipment can be brought to the scene within five minutes this is classed as "rapid attendance" and the ceiling heights can be extended.

Warning devices

Having considered the devices used to trigger the system to operate we must give some thought to the devices which sound or indicate the alarm.

Alarm sounders are devices which give an audible indication of fire in order to alert the occupants of a building. For general guidance, the following points should be observed.

The minimum sound level produced by the sounders is 65 dBA or 5 dBA above background noise level at any occupied location within the building. If persons are likely to be sleeping, for example hotels or living accommodation, this figure is increased to 75 dBA at the bedhead. Where the building is divided into compartments or rooms with doors, at least one sounder per room will be required.

The sound level should not be so high as to create a risk of permanent damage to hearing. It is preferable to have a large number of low powered sounders rather than fewer high-powered devices. All devices within one system should produce the same sound and this must be different from the sound produced by any other warning system.

There must be at least two sounders in any building so that the failure of one sounder does not result in a complete failure of the system.

The recommended frequency of the alarm should be within the range 500 to 1000 Hz. and where a two-tone alarm is used, at least one of the frequencies should lie within this range. An exception to this may be made if the frequency of background noise is between 500 Hz and 1 kHz. In such circumstances a different alarm frequency is acceptable.

Types of sounder

Bells

Figure 4.62   Alarm bell

Modern fire alarm bells are frequently of the motorised underdome variety, available in indoor or outdoor patterns and capable of an output of 90 to 95 dBA at one metre.

Bells are the most common form of fire alarm sounder as they are usually associated with fire and are chosen where other types of sounder are used for different purposes. The wiring is normally terminated on the backplate and the bell is connected by plug-in contacts. This allows quick and easy replacement should it be necessary.

Sirens

Figure 4.63   Siren

The siren is a motorised device in which a slotted rotor rotates within a similarly slotted stator. The sound produced is a distinctive single note of 95 dBA at approximately 800 Hz. Sounders of this type are frequently used to penetrate background industrial noise. However care needs to be taken to ensure that they are not too similar to signals used to start and finish work periods.

Horns

Figure 4.64   Horn

A vibrating diaphragm is used to produce a very penetrating raucous sound. This is used as an alternative to bells or sirens. These devices may have an output of about 100 dBA but this can be moderated by means of a simple adjustment.

Electronic sounders

Figure 4.65   Electronic sounder

These sounders are available in a wide variety of styles. The installer may be able to select a type of output which may vary from single tone, two tone, interrupted (pip-pip) sweep or warbling effect. The frequency or frequencies may be pre-determined or adjustable. This gives the installer a wide range of options to suit the conditions on site, provided that the basic rules are observed.

Visual Indicators

Figure 4.66   Strobe light

This is a specialised application of the rotating beam lamp or xenon strobe lamp.

The unit is situated in a conspicuous position and in the event of an alarm, a very bright flashing light will be emitted. As this will be reflected from all the surrounding surfaces, it is not necessary to be looking in the direction of the lamp in order to be aware of the alarm which should flash 30 to 120 times per minute.

Visual indication is required in areas of high ambient noise levels. They are also used where hearing impaired persons may be present. These visual indicators are only used in addition to audible alarms.

Try this
1. Calculate the number of heat detectors required for a flat roof building 40 m × 15 m.

2. For the same size of building calculate the number of smoke detectors required.

3. If the building is used as a youth hostel dormitory which type of warning device would be suitable?

Zoning, zone indication and supervisory alarms

The zoning of premises is a technique which can help to obtain a fast and effective response in the event of the outbreak of fire.

The following general guidelines should be observed:
1. The floor area of a zone should not exceed 2000 sq. metres.
2. A zone may extend to more than one compartment but these must be complete compartments and the zone bounded by main walls, floor and ceiling.
3. It should not be necessary to travel more than 30 metres to ascertain the position of a fire within a zone.
4. Where detectors are to be situated in stairwells or lift wells these should be regarded as one or more separate zones.
5. Where the total floor area of a building is 300 sq. metres or less, it may be regarded as a single zone even though there is more than one floor.

In a zoned building, a fire call will sound the alarms within that zone and an indication will appear at the fire control panel. Each zone can be evacuated and sealed to prevent the rapid spread of fire. Action can be taken to deal with the fire which is then contained within that zone.

It is essential that clear and unambiguous zone indication is given. A manual or automatic indication of fire should produce the correct response and this is a function of the fire control panel.

Depending on the building's structure and the established fire alarm procedure, it may not be necessary to evacuate more than one zone. Where this is the case, a supervisory signal, usually a light or buzzer, will attract the attention of the person in charge of fire procedure. In these circumstances sounders are grouped to cover the affected zone or zones only. There must also be the facility for sounding a general alarm should this be necessary.

In large buildings, it may be desirable to first evacuate the areas at greatest risk. In these circumstances an evacuation alarm will be given in the high risk area and a fire alert sounded in the rest of the building. The usual arrangement is for continuous ringing to indicate evacuation, and interrupted (on-off) ringing in one second pulses to indicate fire alert.

The system may also be capable of the automatic conversion of the alert to a full evacuation signal if the need arises. Staff and persons using the building must be made aware of the correct procedures to be followed in the event of either of these signals.

In some premises such as cinemas, theatres and hospitals, the management of large numbers of people can present problems with safe evacuation. Panic and confusion can often result in far greater danger than the cause of the original alarm.

Subject to statutory and other requirements, a system of staff alarms such as personal pagers or coded signals, can alert a sufficient number of fully trained personnel who will then organise a safe and orderly evacuation.

Premises having staff alarms must also have means of sounding a full alarm if such action is considered necessary by a responsible person in charge.

Any form of restricted alarm system should incorporate means of summoning the fire-fighting services when the alarm is given.

Automatic communication to the emergency services

Subject to agreement with the fire authority, direct communication links can be set up which will summon the fire service automatically in the event of an alarm.

This may be done by a direct line or through the public telecommunication network.

Figure 4.67   Auto communication

Fire systems which utilise the emergency telephone facility usually incorporate a synthesised or recorded voice message for this task. This informs the emergency services of the nature of the alarm, the premises from which it originated and the best point of access for the fire service. Digital coded systems can transmit the necessary information through the telephone network to suitable decoding apparatus at fire control headquarters along with supplementary information such as "fire", "fault" or "test". In any event, the full co-operation of the fire service and operators of the telephone network is essential if the system is to be efficiently installed and operated.

Cables, wiring and power supplies

The following cables are recommended for use in fire alarm installations subject to any restrictions on their use and the requirement for further protection where necessary.
- Mineral insulated copper sheathed
- Cables complying with BS 6387 such as FP200
- PVC insulated and sheathed singles, twin and multicore
- PVC insulated non-sheathed

- BS 6231 heat resisting switchgear wire
- PVC steel wire armoured cable BS 6346
- XLPE steel wire armoured cable BS 5467
- Co-axial cables to BS 2316 part 3
- Heat detecting cable

Where cables are required to operate during prolonged exposure to fire, they should be one of the first two types in the above list. Other cables may be used but only if they are buried in the structure of the building by the equivalent of at least 12 mm of plaster.

Where cables are at risk of mechanical damage, they should be given adequate protection by any of the normal methods.

Unless otherwise recommended, cables should have a minimum cross-sectional area of 1 sq. mm if solid or 0.5 sq. mm if stranded. Normal cable selection practice will apply and the voltage drop must be kept within a value which will ensure correct operation of the equipment.

If cables are to be run in conduit or trunking, it must be of a type which complies with the fire alarm regulations. The requirements for segregation apply and the appropriate segregation between fire alarm and other circuits must be achieved (Figures 4.68–4.71).

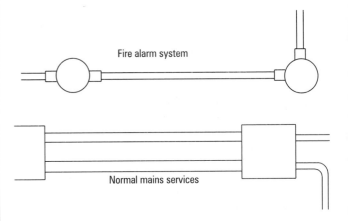

Figure 4.68   Separate conduit or trunking

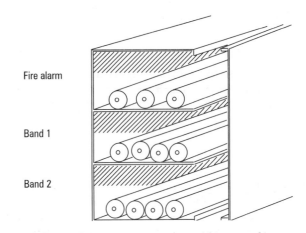

Figure 4.69   Separate compartments within a trunking

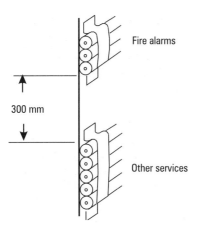

Figure 4.70   Mounting at a distance

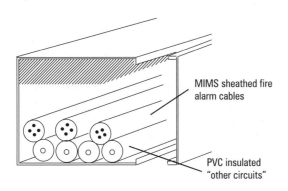

Figure 4.71   Segregation not necessary where MIMS or
cables to BS 6387 are used

Power supplies

For most of the time, the fire alarm main panel and all
auxiliaries will receive their power supply from the mains. A
problem can arise when there is a power supply failure to the
circuit supplying the equipment. To maintain adequate fire
protection there must be some form of back-up system which
is reliable and has sufficient capacity for the purpose.

To try to ensure that the normal supply is as reliable as
possible, it must always be installed in accordance with
BS 7671 and BS 5839. Connection to the mains supply should
be by means of a separate switchfuse and this should be clearly
labelled:

FIRE ALARM: DO NOT SWITCH OFF

Standby supplies

Standby supplies are dealt with in more detail later in this
chapter. However, there are some particular requirements for
fire alarm systems which we shall consider now.

Secondary batteries

A secondary battery, with its own mains-fed charger, is the
usual form of standby supply. It should be of a type suitable for
four years continuous normal service. Lead acid batteries
which have been specifically designed for this purpose are
very reliable. It is most important to ensure that the charging
conditions are suitable for the battery. Frequent overcharging
can adversely affect the life of the battery and undercharging
could result in a failure of the system. The advice of the
manufacturer must be adhered to if the system is to remain
reliable for any length of time.

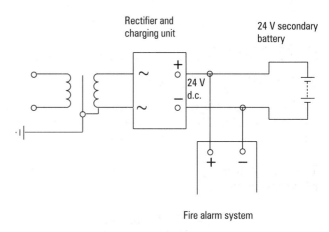

Figure 4.72   Basic standby supply

The purpose of the standby supply is to maintain protection
until the normal supply has been restored or other measures
taken to ensure safety. This requirement is normally satisfied
if the battery has sufficient capacity to maintain the system in
operation for 24 hours. Having maintained the system for that
time the battery should still be able to provide an evacuation
alarm for at least 30 minutes. Certain variations can be applied
to this requirement depending on whether the premises are
supervised or not. For premises left unattended for long
periods, the battery capacity will be increased accordingly.

Operation, care and maintenance of the system

There should always be a suitably competent person who is
prepared to take on the responsibility for the supervision of the
alarm system. The responsible person should lay down the
procedures to be followed in the event of an alarm and should
also ensure that all members of staff are adequately trained.
The responsible person should also ensure that the system is
properly maintained, regularly tested and that liaison exists
between persons whose work might affect the reliability of the
system.

Fire alarm systems need to be regularly maintained and tested
and the responsibility for this lies with the responsible person.
The requirements for the maintenance and testing of the
system are laid down in BS 5839. A log book for the regular
checks of the system, along with records and drawings should
be kept available for inspection. Typical wiring diagrams are
shown in Figures 4.73 and 4.74 and further details of the
requirements for maintenance can be obtained from BS 5839.

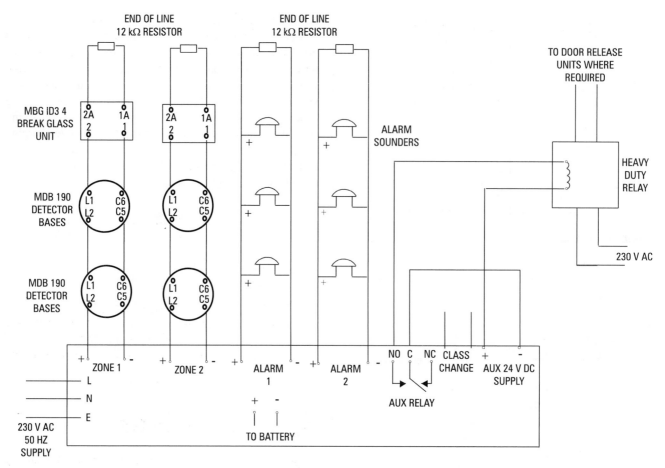

Figure 4.73   *Typical wiring diagram for a fire system control panel*

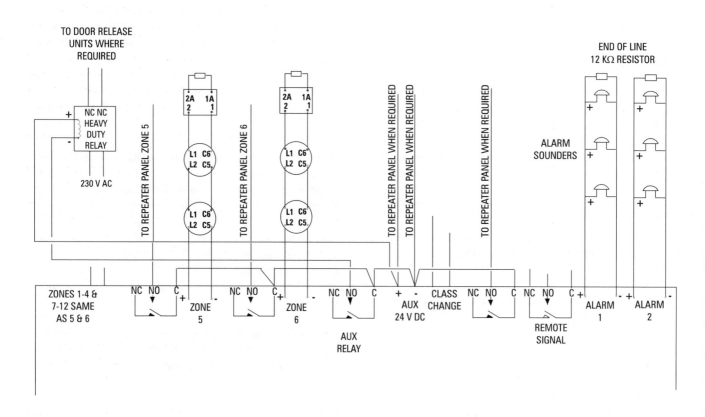

Figure 4.74   *Typical wiring diagram for an addressable fire system control panel*

Self-contained emergency lighting units

Another safety system which we should consider at this time is the provision of emergency lighting. The requirements for emergency lighting systems are covered in BS 5266 and in this book we shall consider the provision of self contained luminaires only.

Where an emergency lighting system is installed, it should come into operation whenever the main supply fails. The principal requirement is that, when the main supply fails to any local lighting circuit, the emergency lighting in the area served by that circuit should come into operation. Self contained luminaires are a simple and effective method of achieving this requirement and they are classified using a simple coding system.

M - Maintained system – In which the luminaire is illuminated continuously irrespective of the source of supply, battery or mains.

NM- Non-maintained – These luminaires will not be lit until a supply failure occurs. The battery will be kept charged from the mains supply but will only be drawn upon to supply the emergency lighting in the event of a mains failure.

Duration – The battery should be capable of providing light output for a continuous period of one or three hours duration, depending on the requirements of the enforcing authority.

Examples:

M/1 Maintained system one hour duration.
NM/3 Non-maintained system, three hours duration.

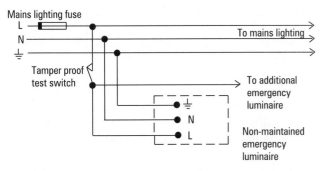

Figure 4.75   Maintained system

Figure 4.76   Non-maintained system

A self-contained luminaire contains its own lamp, battery, automatic charger and change-over device. Self-contained luminaires are available in a wide variety of styles to suit the installed environment and aesthetic requirements. Lamp conversion units can be installed in conventional luminaires in order to provide standby lighting.

The main **advantages** of the self-contained system are;
- ease of installation
- low cost of installation
- independent operation
- flexible (units can be added)
- no need for a separate battery room
- no need for segregated wiring

Disadvantages are as follows;
- short battery life (may need replacement after 5 years)
- ambient temperature range 0 to 25 °C only
- generally does not discriminate between daylight and darkness. May switch itself on and discharge the battery if the supply fails during daylight hours or when premises are unoccupied.

Standby supplies

We have established there is a need for some form of supply for fire alarms and emergency lighting in event of a mains supply failure. There are other "essential services" which may also require a back up supply in the event of such a failure. The mains supply can normally be relied upon and it is easy to assume that it will always be there, readily available and suitable for our needs. It is only when it fails that we begin to realise how much we depend on having a reliable power supply. If the supply should fail then the every day mundane services such as heating, lighting and hot water take on a new significance. Work ceases and tools and machinery come to a standstill and in some instances essential life support equipment may cease to function. With the heavy reliance on electronic equipment for banking, business data, communication and storage then the loss of the power supply to computerised systems can result in loss of vital information. Such loss could have serious financial implications for companies and governments. This loss, risk to life and the cost of lost data and production in many cases is so high that it warrants the provision of a standby supply system to prevent it.

The extent of the alternative supply which may be necessary under "mains failure" conditions will be dependant upon the actual requirements. In a hospital environment there are many areas which will require full service to be available and other areas which may be able to function at a reduced level. This would therefore necessitate a standby supply which is capable of providing a high output for a long period of time.

Computer systems may be designed so that an alarm is sounded on loss of supply. The standby system could then provide sufficient output to allow the system to be closed down without loss of data. Alternatively the company may require a standby supply sufficient to allow essential lighting and full function of the computer system throughout mains failure. The extent of the standby provision will therefore depend upon the extent of the operation which is to be maintained during the mains supply failure.

Let us begin by looking at the principal areas which could require an alternative source in the event of a supply failure.

1. Safety of persons

Hospitals have many areas where the supply is essential for life support such as operating theatres and intensive care units. In addition there are life monitoring systems on many wards and the staff can only continue to maintain care for the patients with lighting and power available. Maintaining the operation of fire alarm and emergency lighting systems, particularly in areas where there may be a risk to life if these services are not available, is essential.

Remember: The hospital standby supply will need to be able to support all the services essential for life support and care of patients and so may be quite a substantial supply. Fire alarm and emergency lighting standby supplies will need to meet the requirements of the appropriate standards.

2. Security of data, goods and premises

The security of data requires a standby supply system and this may be as simple as a battery and inverter for a PC to allow the operator to close down the machine without loss of data. Large organisations dealing in finance or information may require a full standby supply to allow their business transactions to continue without loss during a supply failure. There is also a need to maintain the security surveillance and alarm systems which are installed to protect people and premises. This may also involve the maintenance of supply to security lighting during the main supply failure.

3. Maintenance of manufacturing processes

In many manufacturing industries it is essential to have a supply continuously available. For example a tunnel kiln as used in the manufacture of ceramic products requires an air-flow, not just for combustion but to control the heat distribution within the kiln. If the supply fails and is not restored within a fairly short period, the heat of the material in the kiln will rise, soften the supporting steelwork and the structure will collapse. The cost of such a mishap could run into millions of pounds. In more general terms, the cost of having staff and machines standing idle and the cost in lost production is sufficient incentive for most large manufacturers to invest in standby generating equipment.

4. Changeover and the correct working of equipment

Not only is the continuity of mains supply an important factor in everyday life, but the quality of the supply is also of great importance. It may be quite in order to manually re-start a fan

or pump after the break between mains supply failure and connection to the standby source. This time period may be too long for a computer in the process of transferring millions of pounds from one account to another or for a patient whose next heartbeat depends on the continuity of the power supply.

Maintenance of supplies

The provision of a standby source of power is one thing but to gain access to the standby supply may be quite another. A diesel generator is a very common standby source but its method of connection may vary according to the required speed of response.

Manual changeover

Where speed of response is not essential, a member of staff can go to the "generator shed", start the engine, return to the switchroom and operate the change-over switch. This will then disconnect the load from the supply and connect the essential services to the generator. The installation will then remain in that mode until somebody notices that the mains supply has been restored and the process is reversed.

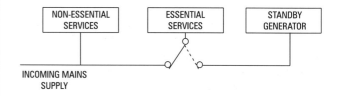

Figure 4.77   Manual change-over

Automatic changeover

Diesel alternator sets can be arranged to start automatically in the event of a mains failure. A monitoring relay will drop out when the mains voltage falls below a pre-determined value.

The engine is started by means of a conventional starter motor which responds to a signal from the mains failure relay. When the engine has attained full speed and has stabilised, a change over contactor disengages the mains supply and connects the load to the alternator. There will be a short break in supply with this process but this may not be more than five to ten seconds. It is inevitable that some contactor controlled loads will have to be manually re-started after a break of this duration. Staff will need to be made aware of their responsibilities following this procedure.

Following restoration of the mains supply, and such time as is necessary to regard the restoration to be permanent, the process will be reversed automatically. The generator will shut down to await a future mains supply interruption.

This arrangement is only effective with proper maintenance and regular testing. The battery for starting the generator must be kept fully charged and checked regularly. The diesel engine must be checked regularly to ensure that all fuel, lubricants and coolant are at their correct levels. Regular checks to ensure that the machine starts and runs at the correct speed under automatic control need to be carried out.

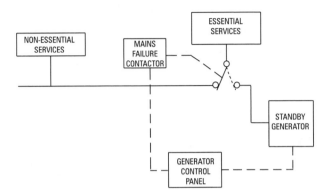

Figure 4.78   *Automatic change-over*

Portable or mobile generating plant

For small power applications a portable petrol generator can provide a supply during the interruption of the main supply. These may be hired from most tool hire companies or because of their modest cost, be bought and kept in storage until required. The output of these generators is normally 240 V single-phase, although 110 V is available for construction site work, and they are available in a range of outputs from about 400 W to 5.5 kW.

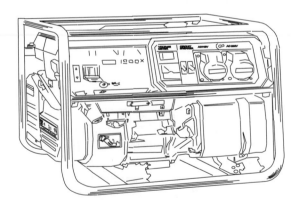

Figure 4.79   *Portable generator*

Battery systems

The main problem which exists with engine driven generators is the break between the mains failure and the connection of the standby supply. This is due to the time taken for the machine to reach its full synchronous frequency and the output voltage to reach and maintain the required level.

Standby battery systems can be brought online with no such break. In a maintained system the battery is already connected to the load and so the loss of the supply has no discernible effect on the load. Non-maintained systems can respond far more quickly than a generator because the battery supply is immediately available and requires no run up time. Changeover is as fast as the switching operation to connect to the load.

For simple, uncomplicated loads such as lighting, a battery standby is often chosen as the preferred system. For computer installations, and other essential services, a battery fed "uninterruptible power supply" (UPS) system is an essential feature.

UPS System

A UPS system provides constant power to the connected load without a break in supply should the main supply fail. This is because the system provides the supply via a static inverter, at all times, both mains and battery supplies. The mains supply is rectified, smoothed and filtered and this d.c. state is the intermediate state of the process. It is supplied to the battery bank to maintain the batteries in a fully charged condition and it is also fed directly to a static converter. This device converts the d.c. input into a sine wave a.c. output which is free from harmonics, spikes and other abnormal components.

During a mains failure, the inverter draws on the battery rather than the rectifier for its supply. A UPS may be incorporated where other sources of standby supply are available, such as diesel generators. The UPS battery system will maintain the supply to the essential equipment whilst the standby supply comes on line. In this configuration the batteries are unlikely to have to supply the load for more than a few minutes at a time. The standby supply will then provide power to the rectifier and hence to the load.

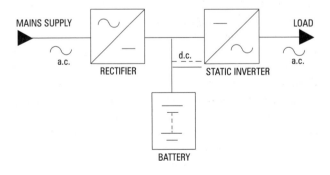

Figure 4.80   *An automatic no-break system*

Types of battery for standby supplies

1. Sealed lead acid gas recombination batteries

These batteries are used where gas discharge during charging
is not permissible. They are compact in size, virtually
maintenance free and can last up to ten years.

2. Tubular plate lead acid batteries

Low internal resistance enables these batteries to deliver high
output currents over a short period making this type
particularly suitable for UPS applications. A large capacity of
electrolyte storage means that as much as 3 years can elapse
between topping-up operations. Battery life can be up to 15
years with good maintenance.

3. Flat plate lead acid batteries

Compact and economically priced, this type of battery is
particularly suited to long standby periods between
discharges. Suitable for long periods without attention and
capable of giving 12 years of service with good maintenance.

4. High performance Planté batteries

The special manufacturing process produces a lead/acid
battery which is extremely reliable and has a life expectancy of
up to 25 years.

If the battery is provided with large reserves of electrolyte,
long periods can elapse between maintenance operations.

5. Nickel-Cadmium batteries

Whereas lead/acid batteries are inclined to suffer if left for
prolonged periods in a discharged state, the Nickel Cadmium
battery has no such problems. It is extremely robust and can
withstand a reasonable amount of abuse, electrical or
mechanical.

They have low water consumption, require little maintenance
and can give good service for up to 25 years.

Central battery systems

The central battery system provides an independent standby
supply which can be used for a variety of purposes which may
include emergency lighting and the maintenance of fire
alarms. Except in the case of very large installations, these are
supplied as self-contained cubicles containing the battery, the
charging equipment and all the necessary control gear.

Fire alarm and emergency lighting circuits must be segregated
from other wiring as required by BS7 671 wiring regulations.

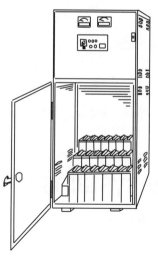

Figure 4.81   Central battery system

For safety reasons, central battery units must be given adequate ventilation to allow for the dispersal of the gases produced during the charging process. These gases are potentially explosive and notices must be displayed in rooms where they are to be housed.

BATTERY ROOM

EXTINGUISH ALL NAKED LIGHTS BEFORE ENTERING

NO SMOKING

The central battery system may be "maintained" in which case the output is continuous and no break occurs on failure of the mains supply. Alternatively, it may be brought into operation by means of a mains failure contactor. To avoid the problems associated with voltage drops over long cable runs in large buildings, it may be preferable to design a system which incorporates several smaller units rather than a few large ones.

Static inverter system

A static converter may be incorporated into a central battery system where non-maintained a.c. supplies are required. These can generally supply the required services for a period of three hours in the event of a mains supply failure. This is described as "passive standby" mode with the inverter being activated only in the event of a supply failure.

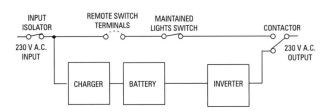

Figure 4.83   Static inverter for non-maintained system

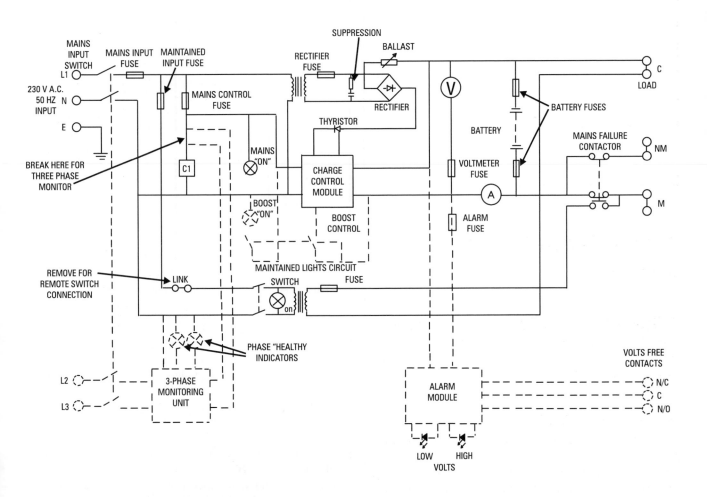

Figure 4.82   Typical central battery unit wiring diagram

The protection of structures from lightning

Our final consideration in this chapter is concerned with the protection of structures against lightning strikes. The main factors that affect the likelihood of a structure receiving a lightning strike are the height of the structure, its location and the degree of exposure. BS 6651 contains the mathematical method of determining the degree of risk for a structure. Where the chances of a lightning strike are high a means of protection should be installed.

In its simplest form, a lightning protection system consists of an air termination, a down conductor and an earth termination. That seems simple enough and in many cases, will be perfectly adequate provided that the system is properly designed and installed.

The design, choice of materials and method of installation are of considerable importance. Particularly when you consider that a lightning strike may discharge a current of some 20 kA into the termination with earth. With an earth electrode of 10 ohms resistance this represents a potential of 200 kV when the current is at its peak.

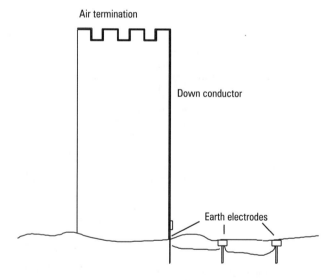

Figure 4.84   Basic lightning protection system

It was considered that a pointed, vertical, copper rod at the highest point of a building would attract lightning away from other parts of the structure. However studies have revealed that horizontal air terminations can be equally as effective.

It is generally accepted that lightning will strike the upper part of a tall building. BS 6651 recommends that an air termination network should be installed which surrounds the roof of the building. If the building has a large flat roof it should also be cross bonded by conductors so that no part of the roof is more than 5m from a conductor. Where there are metal components in the building's structure it is likely that these will be incorporated in the overall lightning protection system. Careful consideration must be given to the way in which these parts are connected and it is essential that appropriate, purpose made clamps and fixing devices are used to make the connections.

Zones of protection

A zone of protection is an area, within which the lightning conductor will attract the strike towards itself.

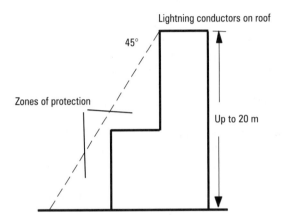

Figure 4.85   Zones of protection (up to 20 m)

For structures up to 20 m in height this is the zone formed within a 45° angle below the roof conductors.

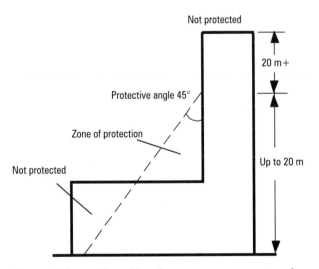

Figure 4.86   Above 20 m there are areas not protected

For structures exceeding 20 m in height, the zone of protection is that which lies outside the circumference of a "rolling sphere" as illustrated in Figure 4.87. For a building of height 100 m BS 6651 recommends a sphere of radius 60 m. Where explosive materials are likely to be present this radius is reduced to 20 m.

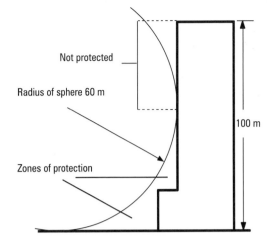

Figure 4.87   Rolling sphere

Remember
During a lightning strike electrical, mechanical and thermal effects place the system under stress.

Where there is a risk of lightning strike to the side of a building, then the air termination network should be extended into that area to provide protection.

Where structures have several different roof levels then it is advisable to have the down conductors from all lower levels joined to the down conductors from the higher parts. This will reduce the risk of side flashing and provide a better overall earth for the lightning current.

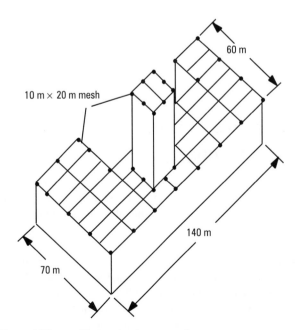

Figure 4.88   Air termination network

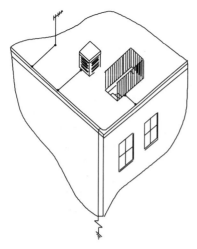

Figure 4.89   Bonding to the air termination network

All projections such as masts, poles and handrails should be bonded to the air termination network.

Try this

Sketch a pattern of conductors which, when bonded together would form an air termination network so that no part of the roof area shown is more than 5 m from a conductor.

Scale 1:1000

Down conductors

The down conductor connects the air termination to the earthing system. Its function is to provide a safe path for the lightning current and to protect the fabric of the building from damage.

Down conductors should be no more than 20 m apart around the perimeter of the roof or the ground level of the building, whichever is the greater. If the building is more than 20 m high, this spacing should be reduced to 10 m between down conductors. In either case, the spacing of the conductors should be symmetrical starting at the corners.

The route should be as direct as possible from the air termination to the earth electrode taking particular care to avoid "side flashing".

Side flashing can occur where the down conductor forms a loop in order to follow the contours of the building.

Where the down conductor follows the outline of a parapet wall, as illustrated in Figure 4.90, the total length of the loop conductor should not exceed eight times the thickness of the obstruction.

To overcome this problem, the conductor may be routed through the parapet – Figure 4.92.

If the building has a cantilevered walkway, the down conductor should not be routed around the inside of the walkway because of the risk of side flashing to persons sheltering inside. The preferred route would be directly to earth as shown in Figure 4.93.

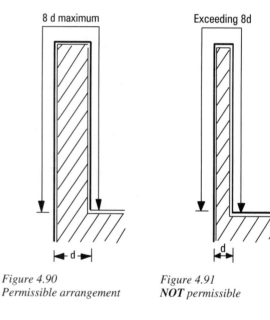

Figure 4.90
Permissible arrangement

Figure 4.91
***NOT** permissible*

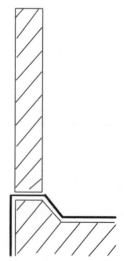

Figure 4.92
Permissible - take the conductor through a parapet wall

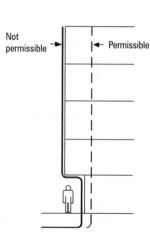

Figure 4.93
Down conductors in a building with cantilevered upper floors

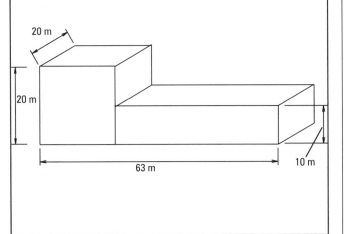

For the duration of the lightning strike the conductors are subject to considerable mechanical forces. The provision of ample, secure fixings is of extreme importance. Fixing devices should be of a type approved by the lightning equipment manufacturer. The devices should be the right size for the conductor and spaced at 1 m intervals to provide suitable support.

A test clamp must be provided to allow for the disconnection of each individual earth electrode for testing purposes. Test clamps should be accessible for testing but not placed in a position which would give access for unauthorised interference.

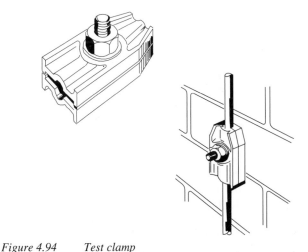

Figure 4.94   Test clamp

The earth termination

This may consist of one or more deep driven earth rods. This is a particularly suitable method because rods can be driven into the soil without the need for expensive excavation work. Providing the rods are of sufficient buried length they will reach soil which is unaffected by seasonal changes. As a general rule, the combined rod length for any one system should be not less than 9 metres. This could be made up of a number of electrodes where there is more than one down conductor. For example two down conductors connected to two 4.5 m electrodes and so on, although it is important to remember that the recommendation is that each electrode should be at least 1.5 m in the ground.

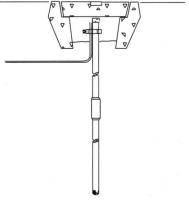

Figure 4.95   Deep driven rod

It is not always possible to drive earth electrodes deep into the ground because of difficult soil conditions. Under these circumstances, several parallel earth electrodes can be driven and connected together with buried copper tape to form a matrix.

The spacing between electrodes should be between one and two times the driven length. At more than twice the driven length there is unlikely to be any improvement in resistance.

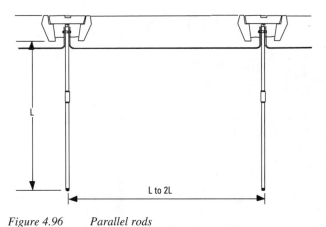

Figure 4.96   Parallel rods

Plate or mat electrodes

Buried plates or mats increase the area of contact with the soil and, as a consequence, lower the resistance of the earth termination. Installing these electrodes is a costly and time consuming process and so this method is normally considered only when driven rods may fail to provide satisfactory results.

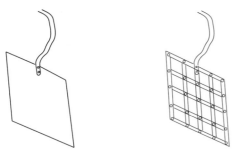

Figure 4.97   Plate electrode   *Figure 4.98   Mat electrode*

Soil conditioning agents

To improve the earth electrode resistance, the plate or mat can be placed in an excavation with special conductive "backfills". These may be commercially known as "Bentonite" or "Maconite" and, being highly conductive, will increase the area in contact with earth. These materials will not harm the earth electrode and will remain effective for the lifetime of the installation. Soil conditioning agents such as salt or other soluble substances are not suitable because they eventually leach out of the soil which will then return to its original condition. There is also the possibility that they will cause corrosion and eventually impair the effectiveness of the electrode.

Bonding

Where earthed metalwork, such as pipes, tanks and structural steel, are located close to the lightning protection system there is a risk that side flashing will occur in the event of a lightning strike. This increases the risk of fire, injury or damage to the structure.

The choice is either to bond or segregate. If sufficient distance can be maintained between these parts they may be regarded as effectively segregated and no further action need be taken. If the distance is too small for safety then they must be securely bonded to the lightning system.

BS 6651 provides a method of calculating the effective isolating distance for such situations. If there is any doubt then the advice of a specialist consultant should be obtained before making a decision.

Remember
The lightning protection system should be bonded to the consumer's earth terminal. Consult your copy of BS 7671 and BS 6651 for the details.

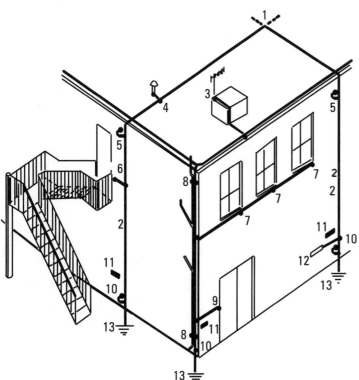

1. Air termination
2. Down conductor
3. Bond to aerial
4. Bond to vent
5. Bond to Reinforcing bar
6. Bond to metal staircase
7. Bond to metal window frame
8. Bond to vent pipe
9. Bond to steel door/frame
10. Test clamp
11. Indicating plate
12. Main earthing terminal of electrical installation
13. Earth termination point

Figure 4.99   Bonding to prevent side-flashing

Inspection and testing

It is recommended that the lightning protection system be inspected and tested at least once in every twelve months. If tests are carried out at a slightly shorter interval then the continuing programme will ensure that all seasonal variations are covered. Where the risk of loss or subsequent danger are high, the inspections and tests should be conducted more frequently.

The inspection and test programme should cover the following points.

1. That the combined earth electrode resistance does not exceed 10 ohms.
2. That all joints in the down conductors are electrically and mechanically sound.
3. That all parts of the system are free from corrosion and securely attached to the structure.
4. That the bonding of the system to the structure and services is effective.
5. That no structural alterations have been made to the building so that the effectiveness of the system is impaired.

Testing

The testing of earth electrodes is covered in a later chapter of this book.

The inspection findings and test results should be recorded in a log book. This should be kept specifically for this purpose and also contain any details of alterations or repairs to the system.

Examples of clamps used in metalwork bonding

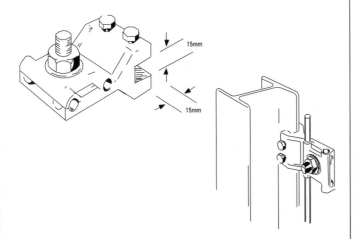

Figure 4.100   Metalwork bond

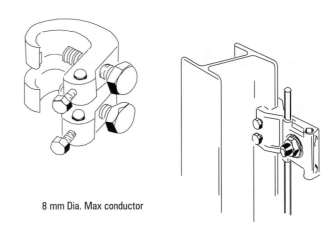

8 mm Dia. Max conductor

Figure 4.101   Re-bar clamp

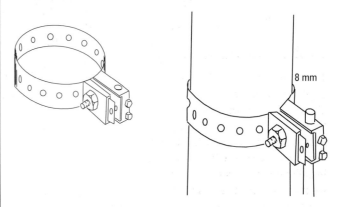

Figure 4.102   Pipe bond

The author and publishers are indebted to Furse (Thomas & Betts) for their permission to use diagrams and information from their consultant's handbook and strongly recommend that all students obtain a copy of this publication as a source of valuable information on the topic of lightning protection

PROJECT

Having completed Chapter 4 of the module you should now consider question 7 of the project.

5

Inspection and Testing

On completion and during the construction of an electrical installation inspection and testing needs to be carried out. As it is the confirmation that the installation meets the requirements of the appropriate standards it is an important part of the electrical work. It may be considered to be the "quality check" on the standard of the installation. It is important to remember that the recording and completion of the records of the inspection and test results has another important role. It provides evidence that the installation was checked and that it was safe to put it into service. The certification you issue also confirms the extent of the installation for which you are responsible. In this chapter we shall be considering the requirements for the inspection and testing of completed and existing installations. Let's start off by looking at the reasons for, and the advantages of, inspecting and testing.

On completion of this chapter you should be able to:

◆ state the purpose and reasons for carrying out inspection and testing
◆ identify items included in the inspection
◆ list in the correct sequence the tests to be carried out in an installation
◆ identify the appropriate instrument required to carry out each test
◆ explain the procedure for each test and verify compliance of given results
◆ enter information and results onto the appropriate forms of certification
◆ state which documentation should be issued to the client following particular work activities

Why inspect and test?

Inspection and testing are carried out to ensure that the installation and equipment:
• meet the requirements laid down in the design
• meet all the requirements of statutory and relevant non-statutory regulations
• are to the appropriate British Standards
• are not damaged in any way so as to give rise to danger
• are installed to the manufacturer's specification

In addition we must consider the statutory implications of the inspection and test requirements. When signing the relevant sections of the forms of certification and reporting, we are taking responsibility for the compliance of that particular aspect with the regulations. So the designer, installer and persons responsible for inspection and testing of the electrical installation have a responsibility to ensure the installation is safe for use.

Remember: Model forms for certification and reporting are contained in Appendix 6 of BS 7671 and Guidance Note 3.

The Electricity at Work Regulations 1989 also places a responsibility on the Duty Holder, that is the person responsible for the electrical installation. It is the Duty Holder's responsibility to ensure the safety of those using the electrical installation. Some indication as to how this is achieved is provided in Regulation 4 which refers to: "As may be necessary to prevent danger, all systems shall be maintained so as to prevent, so far as is reasonably practicable, such danger." The duty holder for the electrical installation may not be electrically trained and therefore unable to make a judgement on the safety of the installation. One method of meeting the statutory obligation is to have a periodic inspection of the electrical installation undertaken. Just having the report does not meet the obligation, and the report should identify any non-compliance with the installation, thus allowing the duty holder to arrange the appropriate remedial action.

It is apparent from these requirements that an electrical installation can, and should, be inspected and tested at several stages during its life. The first such occasion is during construction and upon completion. Inspection is carried out during construction and is an activity that should be carried out by all electricians during the course of their work. This inspection during construction and on completion is essential

to enable the "construction" section of the certificate to be completed and signed. The construction section of the certificate should be signed by the person responsible for the actual construction of the installation. On large sites this section would not be signed by each electrician on the site, it would generally be signed by the person with the overall responsibility for the construction, the site foreman/engineer. This does not remove the electrician's responsibility for inspecting the work carried out to ensure it meets the required standards and specification. On large projects it is not unusual for the client, or the client's consultant, to appoint Clerks of Work who also inspect the installation during its construction and on completion. They however would not be able to sign the construction section of the certificate. Part 7 of BS 7671 contains a list of items which should be included in the inspection process. The wording in BS7671 implies that this list is considered a minimum and that there may be other items which should be included.

So we have established that the electrical installation should be inspected and tested during construction and on completion, but that is not the end of the requirements. Once an electrical installation has been constructed, inspected, tested and placed in service, it should then be regularly inspected and tested throughout its working life. We shall consider the requirements for the inspection of an existing installation before we move on.

Remember

An existing installation is one which has been completed, inspected, tested and put in service. The forms of certification issued for the original installation should be available to the persons carrying out the periodic inspection.

The maximum period between tests depends on the type of premises and its purpose. Guidance on the maximum periods between these periodic inspections is provided in IEE Guidance Note 3.

The list of items included in BS 7671 for initial inspection can also be used for the inspection of an existing installation. It is important to remember that the precise extent of the inspection needs to be discussed with the person requesting the report. There may also be some particular requirements which need to be included for interested third parties, such as licensing authorities. We shall consider the "extent and limitations" of the periodic inspection report later in this chapter. It is as well to remember that each installation is unique and that whilst a list can be quite comprehensive there are always likely to be items which are not covered.

Remember

Any inspection of an electrical installation which requires the removal of covers and accessories should be carried out with the installation, or that part which is being inspected, disconnected from the supply.

Try this

(1) Using IEE Guidance Note 3 complete the following table:

Type of building	Maximum period between Inspection and Testing
Cinemas Theatres Caravan parks Petrol filling stations	
Domestic Places of public entertainment Agricultural and Horticultural Construction sites	

(2) Refer to Part 7 of BS 7671 and locate the check list. Record the regulation number under which this is given and list those items applicable to a domestic lighting circuit.

The purpose of the inspection is to ensure that the installation complies with the required standards and that the equipment installed is fit for its purpose, and environment.

It is a common assumption that the inspection involves simply looking at the installation. Whilst this is part of the requirement, we must also employ the other senses of smell, hearing and touch. For example, a motor which is being overloaded may appear to be OK visually but will give off a distinct smell. It could also feel over-hot to the touch. A loose connection can often be detected by the crackling sound produced as arcing occurs or touching the inside of a fluorescent lighting fitting may reveal moisture and/or corrosion. So it is important to use all our senses during an inspection.

As indicated previously, in the case of a new, alteration or addition to an installation the majority of the inspection is carried out during the construction stage. During a periodic inspection· test however, the inspection of the installation should be carried out with that part of the installation being inspected disconnected from the supply.

Inspection is carried out before testing thereby ensuring that any tests carried out are not invalidated by the later removal of accessories and the like. Remember that it is always possible to inadvertently trap a cable whilst replacing an accessory or fitting. Further information and guidance on the inspection of installations was given in the Stage 1 Design book in this series. We shall consider the main points again here.

Inspection

Cables

Cables need to be inspected to see if they have been properly installed. We need to check for adequate suitable fixings and to ensure that the cable supports are in place. Supports systems and cleats may become loose or damaged by, for example, mechanical impact or corrosion. Terminations, both of the cable and the conductors, should be mechanically and electrically sound. Where cables can be seen, they should be visually inspected for damage from heat, corrosion, mechanical impact and obvious signs of deterioration.

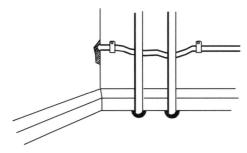

Figure 5.1   Heating pipes have been added which may have not only damaged the cable but may have made it necessary to re-site it.

Try this
Taking the cover off a ceiling rose can expose a multitude of "sins"!

List the "sins" that you can identify in the figure below.

Cable enclosures

Conduit installations need to be checked for different things depending on whether it is steel or PVC conduit. If it is steel conduit corrosion can be a problem, so the conduit should be examined to see if the type of finish is suitable for the environment. When cutting or tightening threads steel conduit can get scour marks which should be cleaned off and painted. A visual inspection may need to be carried out to ensure this has been done.

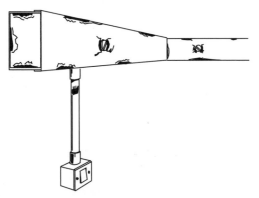

Figure 5.2   *Conduit and trunking may corrode if they do not
have a suitable finish for the environment.*

The fixing of conduit can be a problem because of its own weight and any accessories which are connected to it. An inspection which looks at how well the conduit is fixed to the surface should also consider any likely temperature changes which may result in a change to the characteristics. PVC conduit is particularly prone to temperature change and expansion joints should be fitted to compensate for this. As the conduit is a method of mechanical protection, it should be complete with all covers in place and any unused holes blanked off.

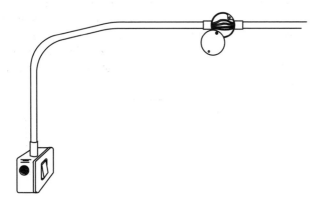

Figure 5.3   *All holes not in use should be plugged and box
lids and the like should be securely fixed in
place.*

Steel conduit, whether it is used as a circuit protective conductor or not, should be connected to the main earthing terminal. PVC and all flexible conduit should contain separate circuit protective conductors. A visual inspection can usually confirm this.

Trunking, both steel and PVC, should be adequately supported throughout its length. Trunking is often fabricated on site and checks need to be made to see if cut edges have been sleeved where cables come into contact with them. As steel trunking has to be electrically continuous, bonding straps may have been required across each joint. These need to be checked to ensure that they are tight. As with conduit the trunking installation should be complete. End caps and lids should be in place and spare holes blanked off.

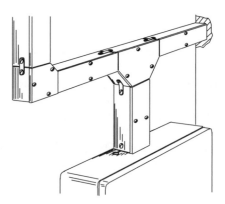

Figure 5.4   *All trunking lids should be in place and coupling
links tight.*

> ### Remember
> Inspection includes using the senses of smell, touch and hearing as well as sight.
>

Switchgear

Switchgear can be visually checked for damage, corrosion and to see if barriers are in place. However, before protective devices can be checked the documentation stating the rating and type of device has to be consulted. It is important to know not only the rating of the device but also the type. For example when BS EN 60898 circuit breakers are used to protect circuits with high inrush or starting currents they will have different characteristics to breakers protecting standard circuits, although both have the same current rating.

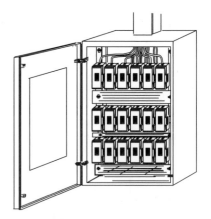

Figure 5.5   *Fuses may need to be checked for the correct
type as well as rating*

Electrical connections need to be checked for tightness and signs of overheating should always be investigated as this may indicate loose connections or overloaded cables.

Accessories

Accessories may contain loose connections as often a number of conductors have to be connected into one terminal. Checks for bad terminations and signs of possible overheating need to be made. In addition to the current carrying conductors the CPC connections should also be checked to see if they are complete between mounting boxes and earth terminals. Damage due to mechanical pressure on the cable insulation should also be considered.

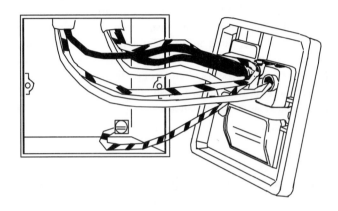

Figure 5.6   Connections should be checked for electrical and mechanical soundness

Remember
An inspection verifies that the workmanship and equipment installed is to the specified standard.

Earthing and bonding

The earthing arrangement at the main intake should be checked to see if it is complete and correctly labelled and all connections and identification labels should be checked. Each circuit protective conductor should be correctly identified by colour coding and terminated relative to the circuits with which they are associated. CPCs also need to be inspected to ensure that conductors are not left bare.

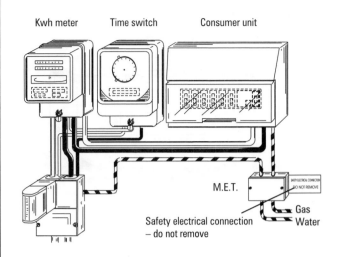

Figure 5.7   Domestic TN-C-S system – main earthing arrangement

In order to use earthed equipotential bonding and automatic disconnection of the supply as our means of protection against indirect contact we need to create an equipotential zone. This requires main equipotential bonding be carried out to all incoming services. A visual inspection to examine connections to gas, water, structural steel, lightning protection systems and the like will need to be made. This inspection should include checks for the correct size, identification and labelling of conductors.

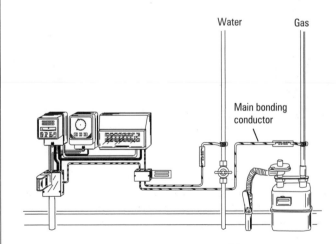

Figure 5.8   Visual inspection should be carried out to confirm connections are satisfactory

Supplementary bonding is required in areas of increased shock risk, such as the domestic bathroom and checks need to be made for good connections and correct identification.

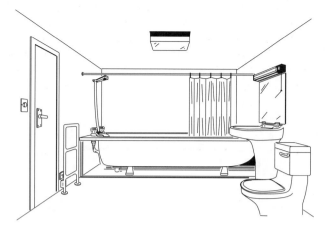

Figure 5.9   *All exposed metalwork should be at the same potential.*
Visual inspections should be made at all connections.

Residual current devices (RCDs) may be used to provide protection against direct or indirect contact where the conditions are particularly onerous or conventional methods cannot achieve the level of protection required. We shall consider the use of RCDs for protection against both forms of contact beginning with indirect contact. During the course of the inspection we should ensure that the appropriate devices have been installed and that they have the correct ratings and settings. It may be possible, at the inspection stage, to check the mechanical operation of the device by operating the test button. The performance of these devices will need to be confirmed as part of the testing process.

Try this
At the point of connection every bonding conductor should have a label.

Write below what this label should state.

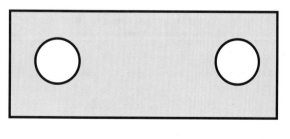

Type of circuit

Inspections also have to relate to the type of circuit used including:
- lighting circuits
- ring final circuits
- circuits supplying bathrooms
- alarm circuits

All circuits require a check to be made to ensure that the correct type and rating of protective device has been installed in accordance with the design. We also need to check that the cable installed is of the correct c.s.a. and current rating for the circuit, bearing in mind the installation method and other relevant factors, such as ambient temperature, grouping, thermal insulation and type of protective device used. However certain types of circuits, or particular locations, have additional requirements which should be considered and we shall look at some of these here.

Lighting circuits

Luminaires should be checked to ensure that they are suitable for the environment in which they are installed, are correctly fitted and, where appropriate, incorporate a suitable size and type of flexible cord. Enclosed fittings, or those where high temperatures are likely to occur due to their operation, should be either supplied in heat resisting cable or have a heat resisting sleeving fitted to the conductors at the terminations.

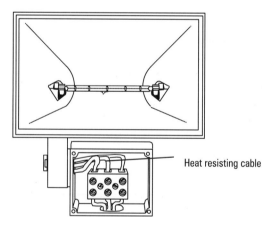

Heat resisting cable

Figure 5.10   *Where high temperatures are experienced, heat resisting cable must be used.*

Ring final circuits

Ring final circuits with socket outlets to BS 1363, are different to other circuits insomuch as they are wired using a conductor whose current carrying capacity is less than the rating of the protective device. This is only possible because the circuit is wired as a ring and effectively has two cables to each point on the circuit, the actual distribution of current around the ring circuit is quite complex and will not be covered in this book. Suffice it to say that BS 7671 does acknowledge the installation of this type of circuit arrangement as being acceptable, providing it is a true ring circuit.

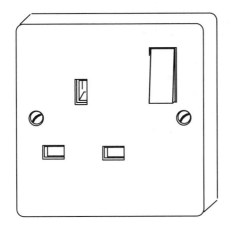

Figure 5.11   13 A socket outlet to BS 1363

If the ring is incomplete, has "multiple loops" (that is a ring within a ring) or has excessive socket outlets supplied via spurs, then currents can flow which exceed the rating of the cables but not that of the protective device. This could result in overheating and damage to the cables and may result in a fire.

During the inspection process we may be able to confirm, visually, that the circuit is a ring where the circuit cables are visible throughout. Where this is not possible, we will need to undertake tests to confirm that the circuit is a true ring. The tests will be considered in the testing section of this chapter.

Bathrooms

Bathrooms are included in the "Special Locations" in Part 6 of BS 7671 and, as virtually every domestic installation has a room containing a fixed bath or shower, are probably the most common location of increased risk encountered in electrical installation work. There are special considerations that have to be made for the type and siting of equipment and accessories within bathrooms and these should be checked during the inspection process.

Figure 5.12   Equipotential bonding leads and connections must be checked

In addition there are particular requirements for supplementary bonding between exposed and extraneous conductive parts within a room containing a fixed bath or shower. The terminations for the supplementary equipotential bonding conductors should be accessible for inspection and testing, so their presence and location should be confirmed during the inspection process.

Having considered the requirements for inspection of the installation, we can look at the requirements for testing.

Testing

Before we can start the testing process we must ensure that we understand the requirements for persons carrying out the tests. We then need to consider one of the most important aspects of the testing process, that is confirming whether a circuit to be worked on is dead.

Persons carrying out tests

The person carrying out the tests should be "competent" and the Health & Safety Executive have defined a "competent" person as:

"A person with enough practical and theoretical knowledge and actual experience to carry out a practical task safely and effectively. The person should have the necessary ability in the particular operation of plant and equipment with which he or she is concerned, an understanding of relevant statutory requirements and an appreciation of the hazards involved. That person should be able to recognise the need for specialist advice, assistance when necessary, and to assess the importance of the results of examinations and tests in the light of their purpose."

The IEE do not give a definition of a competent person. The closest is that for a skilled person which is:

"A person with technical knowledge or sufficient experience to enable him/her to avoid dangers which electricity may create."

practical knowledge

nÿÿ
• theoretical knowledge
and
• actual experience

to carry out the task safely and effectively.

In addition the individual should;
• have the necessary competence in the operation of plant and equipment (including the test instruments) involved in the task

- have an understanding of the statutory and health and safety requirements applicable to the task and the environment
- have an appreciation of the hazards involved
- recognise the need to seek specialist advice or assistance when necessary
- be able to assess the results of examination and tests in connection with their purpose.

To carry out the tests on an electrical installation we shall need an instrument, or instruments, to carry out the range of tests. Section 713 in BS 7671 details the tests required and the sequence in which they are to be carried out. Although eleven separate tests are detailed, it is unlikely that all these tests will be required on any one installation.

Try this

Refer to Section 713 in BS 7671 and familiarise yourself with the tests listed and particularly the different voltage levels involved.

From the information in BS 7671 we can determine that for each test undertaken a voltage will be used. This will be between extra low and high voltage dependant upon the test concerned. In some instances the voltage will be produced by the instrument, in others it is derived from the supply to the installation. In each case it is essential to ensure that the tests are carried out safely. This applies not only to the application of the test, but also to the effect on the installation of the necessary actions required to carry out the tests.

As an example when checking the continuity of main equipotential bonding conductors, they must be disconnected. This means that the installation has no main equipotential bonding during the period of the test. Should a fault develop during this time, extraneous conductive parts could become live. Adequate precautions must be taken to prevent danger for both those carrying out the tests and the users of the installation.

When considering the tests which need to be carried out it becomes apparent that some need to be carried out with the supply disconnected. Others, however, can only be performed once the supply is available. It is important that all the "dead" tests are completed before the supply is connected in order to ensure that undue danger does not arise. If we consider the requirement for correct polarity for example, we should "test" to ensure the polarity is correct before the supply is connected. Once the supply is available we should "confirm" that the polarity is correct. As we shall see later in this chapter, two of the tests we carry out ensure this to be the case.

Try this

For a new installation produce two lists of tests to comprise

(a) those tests which are carried out with the supply disconnected and

(b) those which require the supply to be connected,

Both lists should be in their correct sequence.

(a)

(b)

In addition to the tests identified in Section 713 we may need to carry out a prospective short circuit current test at the main intake position and at the distribution boards of the installation.

Where tests are to be carried out on an installation which is connected to the supply we need to take additional precautions. We need to ensure that the section of the installation to be tested is isolated from the supply in order for the "dead" tests to be carried out. It may be that a voltage is still present when everything appears to be dead. To ensure that the required section of the installation is dead, an isolation procedure, similar to that shown in Figure 5.13, should be adopted.

A voltage indicator, similar to one of those shown in Figure 5.14, is usually most suitable and convenient for the checks required.

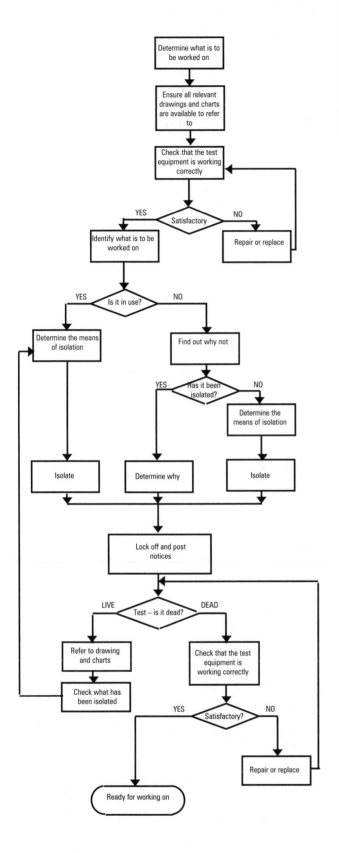

Figure 5.13   Isolation procedure flowchart

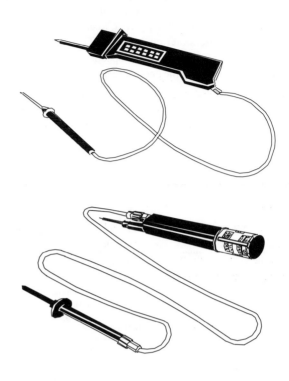

Figure 5.14   Examples of voltage indicators

The Health and Safety Executive produced Guidance Note GS38 which covers "Electrical test equipment for use by electricians". In order to minimise the risk of burns caused by using unsatisfactory test probes they recommend:

- shrouded connectors to the test instrument
- HBC fuses or other current limitation
- finger barriers to guard against inadvertent hand contact
- maximum exposed metal tip of 2 mm.

Instruments

In order to carry out the range of tests the instrument used must be able to perform a number of functions. This can be achieved by the use of a single instrument for each test or multi-function instruments which combine some or all of the requirements in a single unit. It is important to ensure that, whatever instrument, or combination of instruments we use, they meet the requirements of BS 7671 and the relevant safety standards.

Table 5.1 contains a list of the tests which may be required and the properties of the test instrument required in each case. Some of the tests and the associated instruments are specialised and are not required on "standard installations". These tests and the associated instruments are shown shaded in the table. It is not a common practice to carry out the test for protection by barriers and enclosures. This requirement is generally carried out as part of the inspection of the installation to ensure the integrity of the enclosure and that the barrier is in place.

Table 5.1

Tests	Test equipment
Continuity of protective conductors	copper conductor – low resistance ohmmeter (milliohmmeter)
	steel conductor – high current milliohmmeter
Continuity of ring final circuit conductors	low resistance ohmmeter (milliohmmeter)
Insulation resistance	high resistance ohmmeter with d.c. test voltages to suit the installation under test i.e. 250 V, 500 V or 1000 V when loaded with 1 mA. Resistance range needs to be in excess of 1 MΩ
Site applied insulation	high resistance ohmmeter with 4000 V a.c. test voltage applied for 1 minute
Protection by separation of circuits	Test 1 high resistance ohmmeter with 500 V d.c. test voltage
	Test 2 high resistance ohmmeter with 4000 V a.c. test voltage
Protection against direct contact, by barrier or enclosure provided during erection	Test probes complying with IP2X and IP4X, 40 V to 50 V supply and test lamp
Insulation of non-conducting floors and walls	Test 1 high resistance ohmmeter with 500 V d.c. test voltage and electrodes conforming to IEC 364-6-61, 1986
	Test 2 test voltage of 2000 V a.c. and measurement of leakage current up to 1 mA
Polarity	low resistance ohmmeter (milliohmmeter)
Earth fault loop impedance	loop impedance tester
Earth electrode resistance	**either** extra low voltage a.c. supply, voltmeter, ammeter and test electrodes
	or an earth tester and test electrodes
	or in some cases a loop impedance tester may be suitable
Operation of residual current operated devices	RCD tester with suitable ranges

Note: Shaded tests are specialised tests and are not required on standard installations.

Instrument accuracy

Having chosen the correct instruments for our tests we need to ensure that the readings we obtain are as accurate as possible. For some instruments this is simply a case of zeroing the scale with the leads shorted out, others require battery checks to be carried out. It is a fundamental requirement to ensure that the instruments we use are accurate. The instrument manufacturer usually provides some indication of the periods at which the instrument should be calibrated. Calibration is usually carried out by specialist companies and their reference material is normally traceable to a national standard. However the use of instruments on site and their transport around in the company vehicles does make them liable to damage and shock. Regular checks on an instrument should be carried out to check its ongoing accuracy. A proprietary check unit may be purchased for this purpose, alternatively a reference point on a known installation may be used for the live tests, with a set of resistors used for checking continuity and insulation resistance instruments. These checks carried out on a regular basis will ensure that any variation in the instrument will be found early and a minimum of re-testing of installations will be necessary.

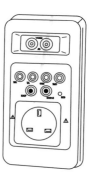

Figure 5.15   Typical check unit

Method for checking ongoing instrument accuracy

Continuity

     selection of 3 resistors; low, middle and high for each range

Insulation resistance

     selection of 3 resistors; low, middle and high for each range

Earth fault loop impedance

     use a reference socket on the company premises, take a number of readings with an accurate instrument and determine the average value. Note this value adjacent to the socket outlet and check the instrument against this value.

RCD

     If an RCD protected outlet is available on the company premises then utilise this in the same way as above, noting the trip times against an accurate instrument. If no RCD socket is available then use the same reference socket on the company premises and a plug-in RCD unit, take a number of readings with an accurate instrument and determine the trip time. Note this value adjacent to the socket outlet and check the instrument against this value.

Taking and recording readings

It is important that we take accurate readings and transfer these onto the record sheets. Where multi-range instruments are used we must ensure that we are using the correct range and reading the corresponding scale. Instruments with digital displays have helped to minimise reading errors but beware of misinterpretation caused by automatic scaling and a "floating decimal point". We must always take care to ensure that the correct readings are taken and recorded.

It is a good idea to record the serial number of the instrument used for each test. This enables us to re-test using the same instrument if necessary. Should an instrument be found to be faulty then all tests carried out can be traced and tests to establish true conditions can be done.

That just leaves us to look at the recording of results before we can begin testing. A number of organisations associated with the electrical installation industry, specialist publishers and instrument manufacturers produce forms for this purpose. Computer programmes may also be purchased which will allow us to record results and either produce the complete form, or print onto proprietary forms. Some form of site recording sheet is normally required, although instruments are produced which have the facility to store and download results. BS 7671 and the IEE Guidance Note 3 provide information and sample sheets for the recording of information and results.

Although there are a number of formats of schedule available there is certain information that has to be recorded. Any schedule that is used must contain as a minimum the information required by BS 7671. A typical schedule of test results form for a single phase installation is shown in Figure 5.16.

Design Details and Record of Test Results

			Instruments used		
				Make	Serial No.
Installation:	Distribution Board Ref:		Continuity		
Contractor:	Earthing Arrangements:		Insulation		
Engineer:	External Impedance (Z_e):		Loop Impedance		
Date of Tests:	Maximum Prospective Short Circuit (I_p):		RCD		
	Supply Voltage				

Circuit Reference	Protection Device			Conductor Size		Continuity					Insulation Resistance MΩ			Polarity (Pass or fail)	Earth fault loop impedance Ω		RCD			
						Radial		Ring											Trip time (ms) at	
	Device (BS)	Type	Rating (A)	Live (mm²)	*cpc (mm²)	Design $R_1 + R_2$ Ω	Measured $R_1 + R_2$ Ω	Design $R_1 + R_2$ Ω	Measured $R_1 + R_2$ Ω		Max. Design	Phase to Neutral	To cpc		Maximum Design Z_s (Ω)	Actual Z_s (Ω)	Rated Trip Current (mA)	50% Y/N	100%	150mA

* IF STEEL TRUNKING OR CONDUIT MARK "ENC"

Figure 5.16

Testing

Now we are aware of the:

- need for inspection and testing
- level of competence required by the person carrying out the inspection and test
- tests required and the suitability of the test instruments
- safety considerations whilst carrying out the tests

it is time to consider the testing.

Remember

When we are carrying out an inspection and test of an installation which has previously been energised and in use we need to ensure that the supply is switched off and the installation or circuit is suitably isolated before we carry out the "dead testing".

When carrying out the inspection and testing on site, it is a good idea to have a copy of the schedules on which to record the results. These can then be used to transfer the information to the final form for issue to the client.

When carrying out testing on an electrical installation the tests should be undertaken in a particular sequence, which ensures that should the installation fail any one test it does not affect the tests already carried out. The IEE has included in their Guidance Note 3, Inspection and Testing, details of the sequence of testing. There are two sequences provided, one for the testing of new installations, the other for periodic testing of installations which have already been in service. The sequences are different to make allowance for the conditions which exist and in each case the sequence should be followed. Simplified versions of the two lists are produced below for ease of reference, but it would be advisable to consult Guidance Note 3 to consider all the information provided within that document.

Sequence of tests for new installations

The following tests should be carried out before the supply is connected (or disconnected as appropriate):

- continuity of protective conductors, main and supplementary bonding
- continuity of ring final circuit conductors
- insulation resistance
- site applied insulation
- protection by separation of circuits
- protection by barriers or enclosures provided during erection
- insulation of non-conducting floors and walls
- polarity
- earth electrode resistance

Once the electrical supply has been connected, recheck the polarity before conducting further tests

- prospective fault current
- earth fault loop impedance
- residual current operated devices

Periodic testing – sequence of tests

- continuity of all protective conductors
- polarity
- earth fault loop impedance
- insulation resistance
- operation of devices for isolation and switching
- operation of residual current devices
- prospective fault currents

Other tests may also be undertaken where appropriate.

For the purpose of this book we shall consider the tests required following the sequence required for a new installation.

Continuity tests

The first tests in the sequence are concerned with confirming the continuity of the protective conductors and ring final circuits. It is important that the continuity of these conductors is confirmed before any further tests are undertaken and we shall consider the tests for each of the protective conductor types and the ring circuit continuity under this general heading.

Continuity test instruments

To carry out the tests for continuity we require an Ohmmeter which has a low ohms range or, as most electricians use, an insulation and continuity test instrument set on the continuity range. Remember we are testing the continuity of conductors intended to carry current and therefore we are likely to obtain fairly low readings, often less than 1 ohm. There are recommendations made in European standards that instruments used for continuity testing should meet the following criteria,

1) an open circuit output voltage of 4 to 24 volts
2) a minimum short circuit current of 200 mA
   (either a.c. or d.c.)

and it would be advisable to ensure that any instrument purchased to carry out continuity tests is manufactured to meet these requirements.

As we are testing quite low values of resistance, it is important that we do not include the resistance of the test leads in our results. We can do this in two ways, the first being to measure the resistance of any test leads before we carry out any tests and then subtract the lead resistance from the result of each test. Alternatively some instruments have the facility to zero the instrument reading to compensate for the resistance of the test leads. The test leads are shorted together and the instrument is set to read zero. Once this has been done, the instrument will automatically read the true resistance of the conductor being measured providing the same leads are used and the instrument is not switched off. The process will need to be repeated each time the instrument is switched on or different leads are used.

Continuity of protective conductors

In order to test the continuity of a protective conductor we must ensure that there are no parallel paths connected which would give a false indication of continuity. For this reason each protective conductor should be tested before it is finally connected to the rest of the installation. On a new installation this would be before the final connection to the distribution board or main earthing terminal. The purpose of the test is to ensure that the conductors are continuous throughout their length and that there are no high resistance joints or connections.

Continuity of earthing and main equipotential bonding conductors

To test the continuity of earthing and bonding conductors we need to employ the use of a long wander lead to test between the main earthing terminal position and the termination at the point where the services enter the building, as these are not usually adjacent to each another. Once the termination has been made to the incoming service we can run out our long lead and carry out the continuity test before we make our final connection to the main earthing terminal.

Example

Test of main equipotential bonding conductor between the main earthing terminal and the incoming water service.

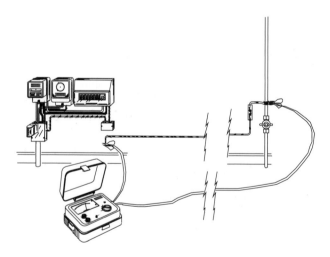

Figure 5.17

Main earthing terminal termination to incoming water
service test value   = 0.5 Ω

Resistance of the test leads   = 0.45 Ω

Resistance of the main bonding conductor

$$= 0.5 - 0.45 \ \Omega \qquad = 0.05 \ \Omega$$

We can verify this value by using the conductor resistance values given in IEE Guidance Note 1, and this gives a resistance of 1.83 mΩ per metre for a 10 mm^2 copper conductor. The length of the main bonding conductor should therefore be in the region of $0.05 \div 0.00183 = 27.3$m and we can verify this by a quick on-site measure.

The requirement to confirm the continuity of protective conductors means we also need to test any supplementary bonding conductors to ensure their continuity. These conductors need to be tested at the point they are installed as they are not connected directly to the main earthing terminal.

Remember

Where supplementary bonding conductors are installed to provide protection against indirect contact the continuity of the supplementary bonds and their connections must be carried out.

Continuity of circuit protective conductors

We can carry out the test for continuity of circuit protective conductors in the same way as we did the main bonding conductors, testing between the cpc at the distribution board and the earthing terminal(s) on the final circuits. However the use of long test leads and running them around the building is both time consuming and a potential tripping hazard for people in the building, and can be particularly awkward in large buildings. The alternative method for this test is to use the associated phase conductor in place of the "wander lead". This method of testing the continuity of the cpc is generally known as the "$R_1 + R_2$ method" as it involves the phase conductor resistance, R_1, and the cpc resistance R_2. We can then record the values for $R_1 + R_2$ onto the schedule of test results. If this method is adopted there is no need to record the value of R_2 separately on the schedule.

C P C continuity using the "$R_1 + R_2$ method"

Whilst this is not essential to establish the continuity of the protective conductor, it is a convenient method and fulfills three of our test requirements in a single operation.
- Confirms continuity of the protective conductor
- Confirms the correct polarity of the circuit under test
- Provides the combined resistance of the circuit earth fault path. This will be required if we are unable to carry out a full earth fault loop impedance test, for example where an RCD is fitted to protect the entire installation.

The method of carrying out this test involves connecting the phase and cpc together, either at the distribution board or at the furthest point of the circuit from the distribution board. The instrument is then connected across the phase and cpc at the other end of the circuit under test.

For the purpose of this explanation we shall consider the use of a shorting link connected at the consumer unit and the test being carried out at the furthest point of the circuit under test. The link used for the connection between phase and cpc should be as short as possible in order to minimise the effect on the readings and where possible the instrument should be "zeroed out" to include the link.

Ensure the supply is securely isolated and insert the shorting link between the phase and the cpc for the circuit to be tested, as shown in Figure 5.18.

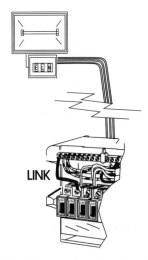

Figure 5.18   *The link is in place in order that the resistance of the phase and circuit protective conductors can be measured.*

The instrument can now be connected between the phase and cpc at any point on the circuit and a reading obtained. However we are not only concerned with the continuity of the cpc but also the "worst case" i.e., where the value of $R_1 + R_2$ is at a maximum, and this is generally at the end of the circuit furthest from the distribution board. The instrument is connected using the crocodile style clips or the probes, whichever is the most convenient, as shown in Figure 5.19.

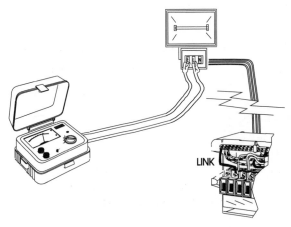

Figure 5.19   *With the link in place the resistance of the cpc can be measured at any point on the circuit.*

Once we are sure that any functional switches on the fixed wiring are closed the test is carried out and reading obtained is recorded and entered on the schedule of test results.

Where a circuit supplies a number of points we can ensure that we have continuity of cpc and correct polarity at each point by carrying out a similar test at each point. The value recorded on the schedule of test results is the highest value obtained, which should be that at the end of the circuit furthest from the distribution board.

Once the continuity testing of the circuit is complete the shorting link is removed and the circuit reinstated.

Remember it is important to test the circuits individually as we must make sure that there are no parallel return paths whilst the test is carried out. The aim of the test is to confirm the continuity of the protective conductor for the circuit and not any fortuitous connections which may not be reliable under fault conditions.

Testing of steel cpcs

Where the cpc for a circuit is provided by a steel enclosure, such as a metal conduit or trunking, the continuity test should be carried out initially by the use of the wander lead or $R_1 + R_2$ method. Should the results or the physical condition of the enclosure give cause for concern over the suitability of the cpc, then a further High Current Test should be carried out. This would need to be done using a high current low impedance ohmmeter. This test will require a supply as the current used is recommended as 1.5 times the design current of the circuit up to a maximum of 25 A, at 50 volts. This test is not usually required on a new installation.

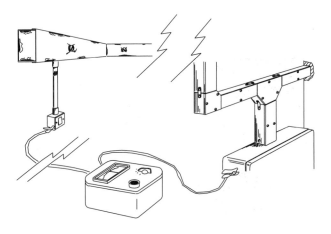

Figure 5.20   *A high current test may be required where steel is used as the circuit protective conductor.*

Continuity of ring final circuit conductors

The ring final circuit has been used extensively for the provision of general purpose power sockets complying with BS 1363, better known as the 13 A socket. However, ring circuits may be used to supply other equipment to overcome particular problems with an installation, location or safety consideration. The method of testing for continuity is applicable to any ring circuit but for ease of explanation we shall be considering a typical BS 1363 ring circuit.

As we discovered earlier in this book, ring final circuits are different to other circuits insomuch as they are wired using a conductor whose current carrying capacity is less than the rating of the protective device. If the ring is incomplete, has "multiple loops", that is a ring within a ring, or has excessive socket outlets supplied via spurs, then currents can flow which exceed the rating of the cables but not that of the protective device. This could result in overheating and damage to the cables and may result in a fire.

Unfortunately it is relatively easy to make a mistake when installing ring circuits, particularly when using single core cables in conduit or trunking. In order to confirm that the circuit is a true ring circuit we need to carry out a sequence of tests on each ring circuit.

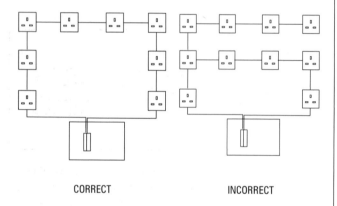

CORRECT   INCORRECT

Figure 5.21   *Line diagrams showing correct and incorrect connections of a ring circuit.*

First we need to ensure that the circuit is securely isolated, all the socket outlets have been installed, no equipment or appliances are connected and disconnect the cables to the ring circuit at the distribution board. This can be done at the origin of the ring in the distribution board, or at a socket outlet close to the distribution board, whichever is the most convenient. As we are considering the initial verification of ring circuit continuity we shall carry out the testing from the distribution board.

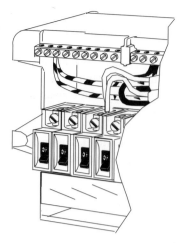

Figure 5.22   *Ring final circuit connected to a consumer unit before tests commence.*
(For clarity the cables to other circuits have not been shown.)

The instrument we use must be a low impedance ohmmeter, or continuity tester, which is capable of measuring small variations in resistance. As we are measuring conductors the values will be quite low and the variations may be small and so the instrument should be able to register values in the order of $0.05\ \Omega$

The first stage of the test is to confirm the conductors are connected in a ring and to establish the resistance of each conductor from end to end. As the ring circuit begins and ends at the same protective device, we simply connect the instrument between the two phase conductors and test to obtain a reading. Remember the resistance of the leads will need to be either zeroed out or deducted from each reading obtained. We shall assume the instrument has been zeroed to take account of the leads. As we are considering the standard ring circuit, the phase conductor will be a 2.5 mm^2 copper conductor which has a resistance in the order of $0.007\ \Omega$ per metre. When we test the phase to phase reading the result, divided by 0.007, will give an approximate total cable length for the ring.

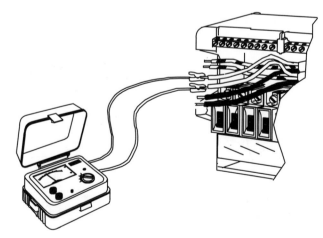

Figure 5.23   *Conductors disconnected from the consumer unit and an ohmmeter connected across the two phase conductors of the ring circuit.*

So if our test result phase to phase on the ring circuit is say, 0.53Ω then we can estimate the total length of the conductor by using the formula

length = total resistance ÷ resistance per metre

in this case 0.53 ÷ 0.007 = 75.7 metres. This length is an approximate value which gives us some idea of the total length of conductor involved.

We now repeat the test with the two neutral conductors and the results should be substantially the same, as the cables should follow the same routes, say within 0.05 Ω.

If the values are not within an acceptable tolerance then the circuit requires some additional investigation to establish the reason, remember that the conductors should follow the same routes and be of approximately equal lengths. If a high resistance is found or an open circuit is established on any of the conductors then, again, further investigation is needed.

We need to record the values we obtain when carrying out this initial test as we shall need them later when confirming the ring circuit continuity. The values are generally recorded as

r_1 = phase to phase resistance

r_2 = cpc to cpc resistance

r_n = neutral to neutral resistance

Providing the tests have confirmed that the conductors have been correctly identified we can proceed with the continuity testing.

The next stage is to "cross connect" the phase and neutral conductors, that is, connect the phase of one end of the ring circuit to the neutral of the other end and vice versa as shown in Figure 5.24. This is relatively easy when the installation is in sheathed cables but may be a little more difficult when using singles in trunking and conduit.

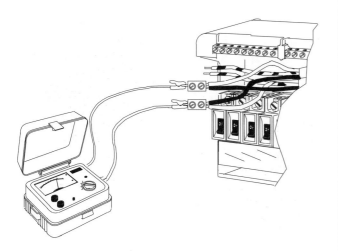

Figure 5.24   *The phase of one cable is connected to the neutral of the other, and this is repeated for the other phase and neutral*

The resistance of the conductors is then measured across the connected pairs as shown in Figure 5.24 and the values obtained are recorded. These values should be approximately half the value obtained for r_1 or when we tested phase and neutral conductors end to end. Now, using a plug top, we can test at each socket outlet between the phase and neutral conductors, with the cross links in place, as shown in Figure 5.25.

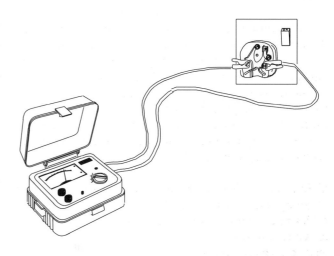

Figure 5.25

Remember
If this test is carried out from a socket outlet, we must also check the reading at the distribution board.

The test readings obtained at each socket and the distribution board should be substantially the same as the reading taken from the test in Figure 5.24. Socket outlets connected via a spur from the ring circuit will give slightly higher values of resistance, and the precise increase will be proportional to the length of the branch cable. If we find that the resistance values increase as we move further away from the distribution board this indicates that we have not cross connected the ring circuit but have connected the phase and neutral conductors of the same ends together. In such cases the ring circuit has to be cross connected correctly and the test carried out to confirm ring circuit continuity and the readings should now be substantially the same at each point.

We must now repeat the process only this time we cross connect the phase and cpc for the ring circuit. The value of resistance will be higher than in the previous test if the cpc is of a smaller c.s.a. than the live conductors. Again we carry out the test at each outlet, between phase and cpc this time, and we should expect some variance in the values. As before the values obtained at sockets connected via a spur will produce higher values, proportional to the length of the branch cable.

117

The highest value obtained during this test process represents the $R_1 + R_2$ value for the ring circuit and should be recorded on the schedule of test results in the $R_1 + R_2$ column. The value obtained should be in the order of $(r_1 + r_2) \div 4$, using the $r_1 + r_2$ values obtained in the first test. This is due to the circuit now comprising of the two conductors connected in parallel and, because they are connected as a ring circuit, the effective length of the conductor is also halved to any point on the circuit, compared to the end to end length. The total resistance value is therefore halved as the effective c.s.a. is doubled and halved again as the effective length is halved, hence a quarter of the original $r_1 + r_2$ value.

Example 1

At the distribution board as in Figure 5.24

Phase to phase	$= 0.8\ \Omega$
Neutral to neutral	$= 0.8\ \Omega$

At the distribution board as in Figure 5.24

$$= 0.4\ \Omega$$

At sockets on the ring when connected as in Figure 5.24

Phase to neutral	$= 0.4\ \Omega$

If all of the sockets are part of a single ring then the reading at each outlet should be substantially the same.

Example 2

For this example we will consider a 25 m ring final circuit wired in 2.5 mm^2 with a cpc of 1.5mm^2.

Phase to phase	$= 0.4\ \Omega$
cpc to cpc	$= 0.6\ \Omega$

At the distribution board, or sockets on the ring, when connected as in Figure 5.25 the resistance should be close to

$$= \frac{0.4 \times 0.6}{0.4 + 0.6}\ \Omega$$

(product over sum)

$$= 0.24\ \Omega$$

Note:

Take care if single-core cables are used that OPPOSITE ends of the phase and neutral conductors are bridged together.

Once we have completed the test, we can remove the cross connections and terminate the ring circuit to the distribution board. We have, by this test process, confirmed the circuit to be a ring, hence the cables will not be subject to overload, confirmed earth continuity of the cpc to each point on the circuit and tested the correct polarity of each socket, three tests in the one process.

Insulation resistance testing

In the previous tests we were concerned with establishing that a good electrical connection was made by recognised conductors and that the resistance of these conductors was sufficiently low for high current to flow in the event of a fault occurring. When we carry out insulation resistance testing, we are testing the resistance of the insulation separating live parts, conductors in particular, from each other and from earth. So as this time we shall be looking for high values of resistance the instrument used must be capable of measuring these high values of resistance, in the range of MΩ.

We are also testing to establish that the insulation is going to be able to withstand the rigours of everyday operation and the voltages likely to be encountered. To test this we use a voltage higher than that to which the insulation will normally be subjected and the test voltage required will be dependant upon the normal operating voltage of the circuit to be tested. Advice on the applied voltage for this test is given in Table 71A of BS 7671 and by reference to this we can see that for circuits other than SELV & PELV and those operating above 500 V the test voltage is 500 volts.

Insulation resistance test instruments

The instrument used to carry out the insulation resistance test must be able to meet the requirements of BS 7671 for the insulation being tested. In this case we are considering installations operating at 230/400 volts and so our test instrument must be capable of producing a test voltage of 500 V. However, as we can see from Table 71A in BS 7671, there are separate criteria for circuits operating at other voltages and instruments used for testing these circuits must meet the appropriate requirements.

The reason for using a higher voltage level is that it is sufficiently high to reveal any breakdown or weakness in the insulation, which is "voltage sensitive". In addition the instruments must be able to supply the test voltage when loaded to 1 mA. This is generally achieved by the use of an instrument with a battery power source and electronic circuitry to produce the required output and may have either an analogue or digital display.

Figure 5.26   Digital insulation resistance tester

Figure 5.27   Analogue insulation resistance tester

Safety whilst carrying out the insulation resistance tests

There are two main aspects to the safe implementation of the insulation resistance test.

The **FIRST** is that the test should be carried out on circuits disconnected from the supply. To assist the test engineer some instrument manufacturers have built voltage indicators into the equipment. If this indicator is activated then no tests should be carried out until the supply has been isolated. Where test equipment does not have this facility, the circuits being tested should be checked to confirm there are no supplies before the insulation resistance tests are carried out. Where circuits contain capacitors the test voltage may charge up the capacitor and care will need to be taken to avoid discharging the capacitor through your body.

The **SECOND** safety consideration that should always be given serious thought is that the test voltages being injected into the installation are high. Precautions should be taken to ensure nobody can become part of the circuit when the test is being carried out. Voltage sensitive equipment should be disconnected or bypassed before tests take place.

Before we can carry out our insulation resistance tests there are a number of procedures that must be carried out.

These include
- ensuring the supply is securely isolated and that there is no supply to the circuits to be tested
- removing all lamps
- disconnecting all equipment that would normally be in use
- disconnecting and bypassing any electronic equipment that would be damaged by the high voltage test, (this may include lamp dimmer switches)
- putting all fuses in place
- putting all switches in the ON position (unless they protect equipment that cannot otherwise be disconnected)
- testing two way or two way and intermediate switched circuits with the switches in each direction unless they are bridged across during the testing

Once we have satisfied ourselves that the above conditions have been complied with, we can begin testing for insulation resistance.

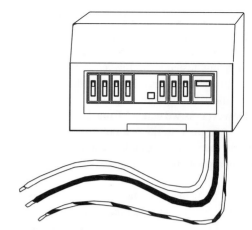

Figure 5.28   *A domestic consumer unit ready for connection to the supply*

In this case the main switch or switches must be on, all fuses complete and in place, or all circuit breakers switched on.

The first test is between phase and neutral as shown in Figure 5.29.

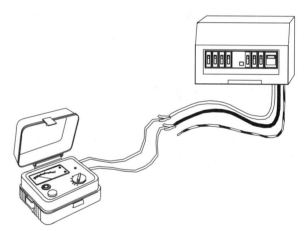

Figure 5.29   *An insulation test instrument connected to the tails of a domestic consumer unit. All switches in the unit are in the ON position.*

The resistance for this should be greater than 0.5 MΩ. If it is less than this further investigation must follow. This can be started by repeating the test with each circuit isolated in turn. When the circuit with the fault has been identified, a further breakdown of this circuit can be made. Often a lamp has been fitted by mistake or the immersion heater is on. However until there is a satisfactory explanation for this result no further tests should be carried out.

When a satisfactory result has been obtained for the first test the phase and neutral can be connected together and the test instrument connected between these and the main circuit protective conductor. As before this resistance must be equal to or greater than 0.5 MΩ. If a meter with an analogue scale is being used, the needle may point to INF or ∞. This means that the reading is greater than the range of the instrument and is not something that can be recorded on a report. If the maximum reading on the meter scale is 100 MΩ then an infinity reading should be recorded as greater than 100 MΩ or > 100 MΩ.

Insulation resistance test on a new installation

We can now begin to undertake the insulation resistance testing on our installation. We shall first consider the process for testing a new domestic installation and then consider the implications of a larger installation, single circuit tests and testing on installations which have previously been energised and in use.

New domestic installation

When testing a new domestic installation this can be carried out from the tails to the consumer unit before they are connected to the supply (Figure 5.28).

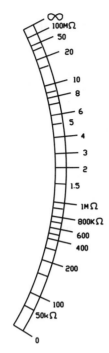

Figure 5.30 The scale of an analogue meter showing "0" at one end and "∞" at the other.

Insulation resistance testing on larger installations

In theory there is very little difference between carrying out an insulation resistance test on a small or large installation. In practice however, there are a number of extra considerations that must be given.

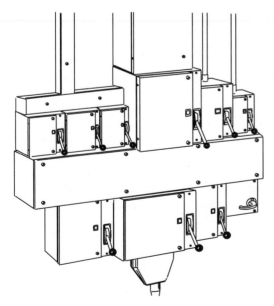

Figure 5.31   A three-phase intake with TP & N bus-bar unit and switch fuses

If all of the preparation was carried out insomuch as the lamps were removed, switches were all on and so on, and a Megohmmeter was connected across the supply cables of a new large installation before the supply was connected, the chances are that the results would not be acceptable. This would usually be due to the enormous total length of cable used in the installation. As the meter is trying to measure insulation resistance and not conductor resistance, the longer the total length of insulation the lower the total insulation resistance would be.

Allowance for this is made in BS 7671 and we can break the installation down into the main switchboard and each distribution circuit tested separately with all its final circuits connected. This will allow us to undertake a test at each sub-distribution board with all the circuits connected and a test at the main distribution position with all the distribution circuits connected but the sub-distribution boards isolated. The principle of this breakdown is illustrated in Figure 5.32 and at each point the total value of insulation resistance must be no less than 0.5 MΩ.

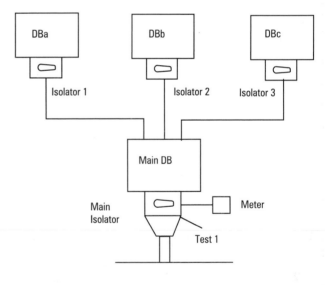

Figure 5.32   Test 1 at the main isolator/Distribution board with Isolators 1, 2 and 3 in the off position

Test 2 at Isolator 1 outgoing terminals with all DB(a) fuses in, lamps removed etc.

Test 3 at Isolator 2 outgoing terminals with all DB(b) fuses in, lamps removed etc.

Test 4 at Isolator 3 outgoing terminals with all DB(c) fuses in, lamps removed etc.

Industrial and commercial installations would usually be designed for use with a three-phase four wire supply. The first test in this case would be between phases. The minimum acceptable insulation resistance is the same as domestic installations, i.e. 0.5 MΩ. After the insulation between phases has been proved to be satisfactory, the three phases can be connected together and the insulation resistance measured between them and the neutral conductor. Once this has been proved to be acceptable, the neutral can be added to the phases

121

and the resistance measured between them and the circuit protective conductors. This again must not be less than 0.5 MΩ. This series of tests may have to be repeated many times to complete the whole installation.

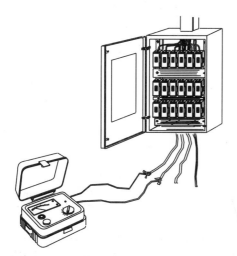

Figure 5.33   An insulation resistance test carried out on a three-phase distribution board

Single circuit testing

When an alteration or addition is carried out to an existing installation, we may need to test that circuit alone in order to compile our certification. We need to make the same checks for equipment and isolation as we do for a full installation and carry out the same procedure for testing the circuit as we would for an installation. We do need to be careful particularly if we are carrying out the test in the vicinity of energised circuits and equipment.

Older and in service installations

On older installations where tests are carried out to verify that the insulation is still in an acceptable condition, it may not be possible to carry out tests on the whole installation at the same time. In these cases each circuit can be tested separately and the results recorded. As equipment is more likely to be in use in existing installations more care has to be taken to ensure it has all been disconnected before the tests are carried out. The acceptable resistances are the same on old installations as on new.

On existing installations sections may have to be isolated and tested to meet the operational needs of those using the installation. The important requirements are that
- all sections of the installation are tested
- a record is made of all test results and to what they apply to. Any results that are not up to specification should be reported in writing to somebody in authority.

Site applied insulation

Carrying out a test for site applied insulation requires the use of high voltage. Care should be taken to ensure all the safety procedures are followed if this test is to be carried out. This involves the testing of insulation which is applied on site and relied upon for protection against direct contact. As the majority of equipment used has the insulation for protection against direct contact provided by the manufacturer of the equipment this test is not normally required. Site applied insulation does not apply to the construction of, say, a control panel using proprietary relays and switches enclosed in a pre-made cabinet using single insulated cables. All the component parts have the insulation applied by the manufacturer before they are supplied to the wholesaler or contractor.

Where the site applied insulation is relied upon for protection against indirect contact then the test should confirm that the insulating enclosure provides a degree of protection equivalent to a least IP2X or IPXXB. These degrees of protection will be considered further in the requirements for "protection by barriers and enclosures" provided during erection, later in this chapter.

The test involves, besides the insulation resistance test, the application of a high voltage test, where the equipment is subject to a flashover or breakdown test at an equivalent voltage to that for a manufacturer's type test for similar equipment. As this test requires specialist equipment and knowledge, we shall not be considering the requirements in this book. Should you be required to have equipment tested in this way it would be advisable to contact a specialist in this field of work.

Protection by separation of circuits

This test is required where the method of protection is by the use of SELV or PELV and electrical separation. For these systems to be employed and provide protection against electric shock the source of the supply must provide separation from live parts of other systems. A simple example of electrical separation is an electric shaver socket manufactured to incorporate a BS 3535 isolating transformer. This, by nature of its construction, provides an output of 240 volts which is electrically isolated from the supply completely. This item of equipment provides protection for the single outlet incorporated within it and uses electrical separation to provide protection against electric shock.

The protection by electrical separation requirements for circuits extend beyond this to incorporate the source of the supply and any associated wiring.

Sources of supply

There are a number of sources of supply detailed in BS 7671 and the most common, for general installation work, are those derived through a BS 3535 isolating transformer. It is usual for such sources of supply to be type tested by the manufacturer and provided it can be shown that the source has been type tested and complies with the appropriate standard then no further testing of the source will be necessary. If this is not possible then it will be necessary to carry out tests to confirm that the source of supply does meet the standard. As these tests are highly specialised, we shall not be considering them here and should they become necessary it would be advisable to contact an expert in this field.

To verify that the output is completely separate from earth an insulation test must be carried out. This test is carried out at 500 V d.c. for one minute and at the end of that period the resistance should not be less than 5 MΩ.

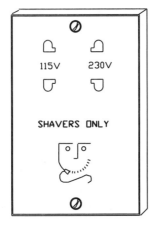

Figure 5.34   *A shaver socket to BS 3535 incorporating a double wound transformer with no connection to earth on the secondary winding*

Protection by barriers or enclosures

So far all of the tests have involved the use of instruments to electrically test the soundness of the installation. It is however possible to have the situation where equipment can pass the electrical tests but still leave live parts exposed to touch.

In order to ensure this is not the case we need to carry out tests to ensure that there is no possibility of contact with live parts. Such risk may arise as a result from the construction of an item of equipment or enclosure on site. More often it is the result of modifications made to items of equipment, such as making entry holes into consumer units and accessories, in order to take our cables into them. There are two principal tests which need to be undertaken and both refer to the requirements identified within the IP code.

The first requirement is that all barriers and enclosures must meet the requirements of IP 2X or IP XXB, and this involves the use of the "standard finger" test probe, as shown in Figure 5.35.

The "standard test finger" is capable of bending through 90° twice, as a normal finger does, and is intended to establish whether there is any possibility of contact with live parts through the insertion of a finger, without the risk of electric

shock to the person carrying out the test. In terms of contact the requirement of IP XXB is the same as IP 2X.

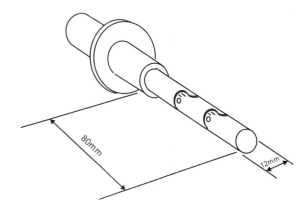

Figure 5.35   *A test probe to IP2.*
Joints may be used to simulate a finger.
For full details and dimensions refer to BS 3042.

The second test is applied to the top surfaces of enclosures and is more onerous than that required for IP 2X. For this test any opening on the top surface of the enclosure must offer protection against the entry of a wire or solid object larger than 1 mm in diameter, protection to IP4X.

Should any of the openings fail to meet the requirements then the opening must be reduced to do so. Providing the openings are suitably sized, and any unused openings are closed, a visual inspection will generally be sufficient to ensure the installation meets the requirements.

Insulation of non-conducting walls and floors

The use of this method of protection is restricted to certain special locations, and the circumstances are such that the installation needs to be under effective supervision. As this type of installation is carried out by specialist contractors and requires specialist testing procedures we shall not consider them in this book. Further information and guidance on this subject is contained in IEE Guidance Note 3, Inspection and Testing.

Polarity testing

We have considered the testing of polarity when we were carrying out the $R_1 + R_2$ earlier in this book and it was stated that we could use this to confirm the polarity of circuits being tested. We do have to ensure that all the circuits are correctly connected and that all protective and control devices are connected in the phase conductors. In order to do so we shall consider the process necessary to confirm this which must be carried out before the installation is connected to the supply.

For safety reasons all switches, overcurrent protection devices and control contacts, must be in the phase conductor (Figure 5.36).

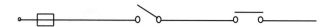

Figure 5.36   All devices capable of breaking the circuit
               should be placed in the non-earthed conductor,
               i.e. the phase

The only exceptions to this are where phase and neutral conductors are operated simultaneously with mechanical links coupling the operation.

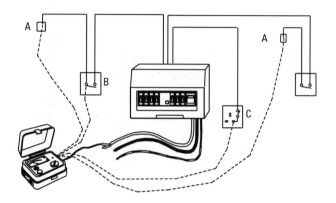

Figure 5.37   Polarity tests are carried out to verify that
               A – the phase at the lamp holder is switched
               B – the switch is in the phase conductor
               C – the switch to the outlet is in the phase
               conductor

Polarity tests can be carried out using continuity test instruments. There is no requirement to record readings as this is a check to ensure that the installation connections have been made in the correct conductors.

On domestic installations the tests should verify that:
- all lighting switches are connected in the phase conductor
- the switches on switched socket outlets are connected in the phase conductor
- the correct pin of socket outlets is connected to the phase conductor
- where double pole switches are used, such as on immersion heaters, the phase and neutrals have not been swapped over

Industrial installations will include all those listed for domestic installations but will also include others due to the complexity of the installations. Most commercial and industrial installations are connected to a three-phase supply and include some three-phase loads. Control devices and isolators used in these circuits must switch the three phases simultaneously so that a single phase cannot be left connected.

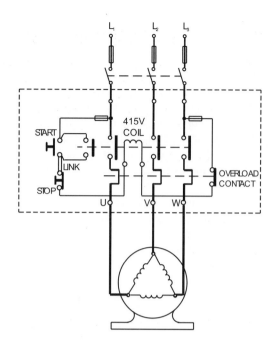

Figure 5.38   Three-phase direct-on-line starter with a triple
               pole switch fuse

So far we have considered tests that need to be carried out on the installation to ensure it is in good condition and that it is safe for the supply to be connected. Once we are satisfied that the installation is safe to energise we can connect the supply but before the installation can be put into service we must carry out some further tests and checks. The following tests are carried out once the supply is available and the installation has been confirmed as safe to energise.

Try this
When carrying out the $R_1 + R_2$ test for continuity of protective conductors the test covers the requirements of which three separate tests which would otherwise need to be undertaken?

Earth fault loop impedance test

A designer, when calculating out an installation, should confirm at each stage that, should a phase to earth fault develop, the protection device will operate safely and before any permanent damage can occur in the installation. The time protection devices take to operate is directly related to the impedance of the earth fault path. Although the designer can calculate this in theory it is not until the installation is complete that the calculations can be checked.

Remember this test, unlike the others we have looked at so far, must be carried out when the installation is connected to the Supply Company's cables. The instrument used for this test is an Earth Fault Loop Impedance Test instrument, sometimes referred to as a "Loop Tester".

External earth fault loop impedance Z_e

The first earth loop impedance test we need to undertake is to establish the value of Z_e. This is the earth loop impedance external to the installation on the supply authority's equipment. When considering TT systems there is no earth path provided by the supplier and so we need to measure electrode resistance, which is covered later in this chapter. To measure the value of Z_e we need to disconnect the installation protective conductors and any bonding conductors from the main earthing terminal. In any event the installation will not be connected to earth throughout this test and so the whole installation must be isolated in the interests of safety.

The test is carried out between the incoming supply terminals and the main earthing terminal (MET). This is generally carried out at the incoming terminals of the main isolator and the MET as shown in Figure 5.39. The value we obtain carrying out this test should be recorded on the certificate as the Z_e for the installation. As a check you should obtain the maximum value for the type of system from the Public Electricity Supplier and the measured value should be no greater than the maximum given. If this is not the case the PES will need to be informed and action taken to resolve the matter before the installation is placed in service. Of course the PES may not, in the event, be able to provide an earth and the installation would then need to be made TT. In the majority of cases the measured value is less than the maximum, often by quite a considerable margin.

Once the value of Z_e has been determined the installation earthing and bonding conductors should be reconnected to the MET, **BEFORE the installation is energised**.

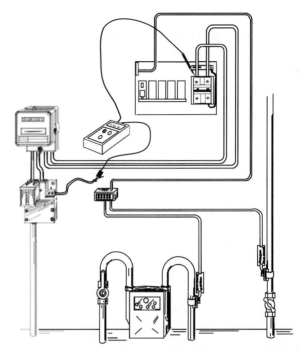

Figure 5.39 *Test between the incoming supply terminals and the earthing conductor*

Prospective fault current

Once we have measured Z_e **and** reconnected the earthing and bonding conductors, we should also establish the prospective fault current at the origin of the installation. We need to establish the prospective fault current, which we shall call I_{pf}, to ensure that the protective devices can withstand such currents.

NOTE:
BS EN 60898 supersedes BS 3871 for circuit breakers. Under the BS ratings were given as "M" values e.g. M6 was a 6 kA maximum fault current, M9 was a 9 kA. This being the maximum value of fault current for that particular type of device.

There are instruments available which are able to provide a direct measurement of I_{pf} but BS 7671 does allow for the value to be determined by calculation or enquiry. The result of an enquiry is generally the maximum value from the Public Electricity Supplier, and this does not really reflect the actual value present, so we shall cover the measurement and calculation of I_{pf} here.

What is prospective fault current?
For our purposes I_{pf} is the maximum fault current that will flow at any point on the installation if a fault having zero impedance occurs between conductors or between conductors and earth. The maximum value for the installation will be at the supply intake position – between which conductors this maximum fault current occurs will depend upon the type of system involved.

Where is the maximum value likely to occur?

The maximum value of I_{pf} is likely to occur at the main intake position and we shall need to determine the maximum value to record on the certificate. In general terms the maximum values will be found:

Single phase supply: TN or TT System Phase to Neutral

Three-phase supply: TN or TT System Phase to Phase

However, on TN-C-S systems, with all the bonding and protective conductors connected to the MET, it is quite possible that the Phase to Earth value for a single phase supply is the highest.

How do we determine I_{pf}?

Measurement

Where an earth fault loop impedance instrument is capable of providing a direct measurement of I_{pf} then the connections are made between phase and earth and phase and neutral as shown in Figures 5.40a & 5.40b. This is carried out with all the earthing and bonding conductors connected to the MET and the instrument set to read fault current. **Note:** When measuring P-N with a 3 lead instrument it will be necessary to connect both the neutral and earth leads to the neutral terminal or the instrument will read the P-E value. The largest of these two values is the one to be recorded on the certificate and for which the protective devices used must be rated.

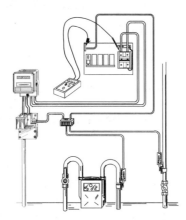

Figure 5.40a   Connections between phase and earth

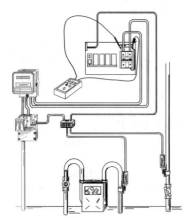

Figure 5.40b   Connections between phase and neutral

Calculation

Where the earth fault loop impedance test instrument does not have a direct current reading we must establish the I_{pf} through the measurement of impedance and calculation using Ohms Law. The impedance values are measured in the same way as the direct current readings. The only difference is that we shall obtain a value in Ohms. Once we have established the impedance values we can then determine the maximum I_{pf} by dividing the measured value of voltage, P-N and P-E by the respective impedance value. So if the P-N impedance was 0.04 ohms and the P-N voltage was 230 V then

$$I = \frac{230}{0.04}$$

$$= 5750 \text{ A or } 5.75 \text{ kA.}$$

If the P-E impedance is 0.06 and the P-E voltage 230 V then

$$I = \frac{230}{0.06}$$

$$= 3.83 \text{kA.}$$

So the maximum I_{pf} would be 5.75 kA.

Three-phase supplies

Where a three-phase supply is provided the highest I_{pf} will be between phases at the origin of the installation.

Warning: Unless the test instrument is specifically designed to be connected to 400 V NEVER connect between phases to measure impedance or current. Only if the instrument is designed for the purpose should actual measurements be taken and the manufacturer's instructions should be followed at all times.

The calculation of the I_{pf} between phases can be determined by calculation using the P-N values. Most supply cables have equally sized phase and neutral conductors so the impedance value P-P should be the same as the P-N value. We could determine the I_{pf} by the use of the formula

$$I_L = \sqrt{3} \times I_p$$

However as a "rule of thumb" we can determine I_{pf} by

$$2 \times I_p$$

so in our above example the P-P I_{pf} would be

$$2 \times 5.75 \text{ kA} = 11.5 \text{ kA.}$$

(Using the $\sqrt{3}$ value the answer would be in the order of 10 kA).

The simple rule of thumb calculation can be readily carried out and always provides a margin of safety, being slightly higher.

If a problem with compliance is encountered using this figure then a more detailed calculation would need to be undertaken.

Testing earth fault loop impedance Z_s

Where a circuit is protected by an RCD a standard earth fault loop impedance test instrument will cause the device to operate. Some instruments are available which can test circuits with RCDs but these carry out the tests using a different test procedure and often a different test voltage or current. Where the instrument does not have this facility the means of determining Z_s is to use the value of $R_1 + R_2$ for the circuit involved and add that to the value of Z_e for the installation. If the circuit is supplied via a distribution circuit we should use the value of earth loop impedance measured upstream of the RCD, Z_{db}, and add the value of $R_1 + R_2$ to that value. So $Z_s = Z_{db} + (R_1 + R_2)$

A similar "tripping" problem can be encountered with some circuit breakers, such as 6 A Type B MCBs, due to the level of current used for the test. Where such devices are used the value of Z_s should be determined by $Z_s = Z_{db} + (R_1 + R_2)$

Remember
Where it is not possible to make a direct reading of earth fault loop impedance, then Z_s is derived by
- measuring the earth fault loop impedance on the supply side (upstream) of the RCD or breaker
- measuring the $R_1 + R_2$ of the circuit downstream of the device, at the furthest point on the circuit from the supply
- adding the two values together to determine the overall Z_s of the circuit.

When we determine the overall Z_s of a circuit in this way we need to make a note on the certification to inform the recipient as to how the value was measured.

Testing for earth fault loop impedance can be made from any point on the circuit but the value which is recorded is the highest value on the circuit. This is the worst case scenario for the circuit where the loop impedance is at its highest and the fault current will be at its lowest. Such conditions are generally found at the point on the circuit furthest from the supply.

Testing earth fault loop impedance on BS 1363 socket outlet circuits is quite straightforward as a lead is provided with the instrument, complete with plug. This is simply inserted in the socket in order for the test to be carried out. The instrument carries out some initial checks
- is the polarity of the connection correct and
- is a suitable connection to earth available?

If the circuit under test fails either of these checks then the necessary remedial work must be carried out before testing can begin. When the instrument indicates that all is well the test can proceed and the reading taken.

The instrument injects a current of about 25 amperes through the circuit under test, from the earth connection around the fault path shown in Figure 5.42, and back via the phase connection. Whilst this test is being carried out the potential on the exposed and extraneous conductive parts throughout the installation will be at a potential above earth potential. If the circuit is incomplete and the test is undertaken then there is a real risk of electric shock if someone inadvertently becomes part of that circuit. Equally if a person forms a connection between the installation earthing system and exposed conductive parts and true earth there is also the risk of electric shock. It is important that we ensure the circuit is complete and warning notices are placed to advise anyone in the vicinity that testing is being carried out and that the installation should not be used.

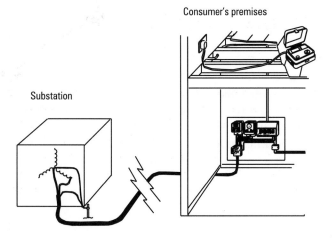

Consumer's premises

Substation

Figure 5.41   *An earth fault loop impedance test carried out on an installation. The substation supplies the installation via an underground cable.*

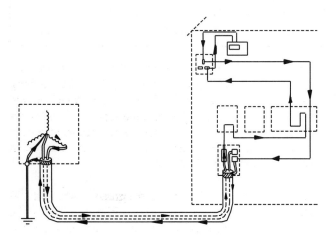

Figure 5.42

Testing circuits which supply equipment other than socket outlets follows the same procedure but requires the use of a set of proprietary test leads, as shown in Figure 5.43, to make the connection to the circuit.

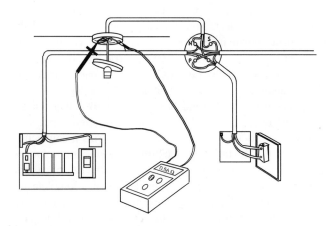

Figure 5.43   *A set of proprietary test leads connected to a suitable light fitting*

Again the value we record should be the highest obtained on each circuit.

The test can be taken from any supply point and should be carried out on all types of circuit. A socket outlet circuit is straightforward as the test equipment just has to be plugged in.

Once we have recorded the values of earth fault loop impedance we need to check these against the maximum acceptable test values for the type of protective device used. The values we considered earlier in Tables 41 from BS 7671 are the design values for the circuit and these values are calculated for when the circuit is operating under normal conditions, i.e. carrying load current. When we carry out our test the circuit is not loaded and therefore the conductor is operating at a much lower temperature than it would be on load. Also the fault current which flows will produce a further rise in temperature which our test will not be able to simulate. For these reasons we cannot use the maximum values of Z_s given in the tables. As a rule of thumb the maximum values of Z_s for test results should not exceed 80% of the values given in BS 7671.

For example if the protective device is a BS EN 60898 32 A type B circuit breaker the maximum design Z_s from table 41B2 is 1.5 Ω. The maximum value acceptable as a test figure is going to be 80% of $1.5 = 1.5 \times 80 \div 100 = 1.2\ \Omega$.

When we check the results of our test they should not exceed the modified maximum test value, calculated by the rule of thumb above. Should the values exceed the maximum then further investigation is necessary to establish the cause and remedy the situation before the installation can be put into service.

Polarity checks

Having tested for polarity prior to the supply being connected we need to test the installation to verify that the polarity is in fact correct. This may be confirmed whilst carrying out the earth fault loop impedance tests.

Earth electrode resistance test

Every installation that forms part of a TT system will have its own earth electrode forming part of the earth fault return path to the supply company's transformer. We need to test the resistance of this electrode to the general mass of earth to ensure that the installation meets the requirements of BS 7671. There are a number of ways to carry out this test and to some extent it will depend upon the location of the installation as to which is the most appropriate test to use. We shall consider the two most common methods of testing the earth electrode. The first being the use of an earth fault loop impedance test instrument. The second being a proprietary earth electrode test instrument.

For safety, as the test calls for the electrode to be disconnected from the main earth terminal, both methods require the installation to be isolated from the supply before the test procedure is started. It is preferable to disconnect the electrode from the earthing conductor, but where this is not possible the earthing conductor may be disconnected from the main earthing terminal.

Earth fault loop impedance test instrument method

As this method requires the use of the incoming supply to provide the power for the instrument it may prove to be advantageous to isolate the installation from the supply, disconnect the earthing conductor from the main earthing terminal and connect the instrument as shown in Figure 5.44.

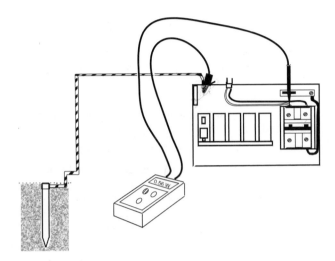

Figure 5.44

Once the earth electrode is isolated from the installation main earthing terminal the earth electrode resistance can be measured in the same way as we measured Z_e for the TN installations. In this case, however, as we are testing the resistance of the electrode to the earth and the return path through the mass of earth to the supply transformer electrode and back, via the phase conductors to the installation, the reading on the instrument is likely to be much higher than that recorded on the TN systems.

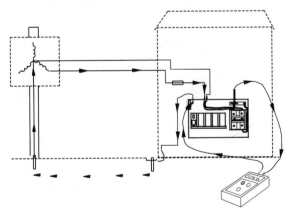

Figure 5.45   *Testing to the earth electrode measuring the earth fault path which gives an indication of the earth electrode resistance.*
To simplify the drawing the supply company's main equipment has been omitted.

Proprietary earth electrode test instrument method

This method involves the use of a proprietary test instrument, specifically designed for the purpose. There are two main types of instrument for this, one having three terminals the other having four terminals. The four terminal type requires two of the terminals to be linked together and you should refer to the particular manufacturer's instructions for this. The remainder of the requirements are the same for both tests and so we shall consider the test instrument with three terminals for ease of reference.

Remember
The termination references for the instrument may vary from one manufacturer to another so always check the instructions to ensure the connections are correctly made before testing begins.

For this test it is advantageous to make the connection directly to the earth electrode and where possible the earthing conductor should be disconnected from the electrode once the installation has been isolated from the supply. If this is not possible the earthing conductor should be disconnected from the main earthing terminal and isolated. The test should then be carried out at the earth electrode.

A second electrode is driven into the ground some 20 to 30 metres from the installation electrode which is under test. A third electrode is then driven in halfway between the two electrodes as shown in Figure 5.46.

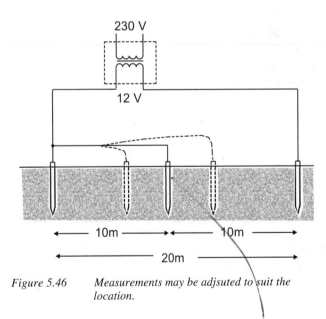

Figure 5.46   *Measurements may be adjsuted to suit the location.*

The three terminal instrument is then connected with the terminal marked "E" to the installation electrode, the terminal marked "C" to the electrode furthest from the installation electrode and the terminal marked "P" to the central electrode, as shown in Figure 5.47, using separate leads for each electrode.

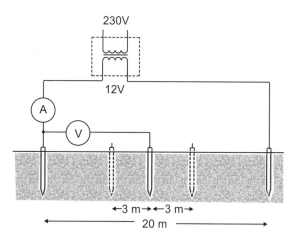

230V

12V

A

V

←3 m→←3 m→

20 m

Figure 5.47

The principle of the test is that the instrument passes current between the two outside electrodes and measures the voltage drop across the connection to earth of the installation electrode and the centre electrode. This provides enough information for the instrument to measure and record the resistance of the electrode, in ohms, on the instrument scale. This test is repeated with the centre electrode approximately 3 m closer to and 3 m further away from the installation electrode. If these three tests produce results that are substantially the same, say within about 5%, then an average of the results should be calculated and recorded as the electrode resistance for the installation.

If the results are not substantially in agreement it indicates that the resistance areas of the electrodes are overlapping. The furthest electrode from the installation electrode must be moved further away and the tests repeated as before.

The main drawback to this method of testing is that it requires a considerable amount of open space adjacent to the installation. This space needs to be suitable for the installation of the test electrodes at the three locations, so metalled car parks, footpaths and the like are not suitable. For this reason the locations at which this method of testing electrode resistance can be used are relatively few. The test is generally carried out using an earth fault loop impedance test instrument.

Testing residual current devices

There are many occasions where RCDs are used to provide additional protection against electric shock from direct or indirect contact. Wherever such a device is installed we must test to ensure that disconnection is achieved within the time required to provide protection from electric shock.

As we established earlier, the two most common ratings for RCDs are 100 mA and 30 mA, and both types require testing to ensure correct operation. There are two tests which need to be carried out on each device operation, when the periodic test button is operated, and the simulation of a fault by using an RCD test instrument. It is important that the electrical tests are carried out before the test button is checked as this may effect the performance of the device.

The RCD test instrument needs to supply a range of test currents appropriate to the device being tested and display the time taken for the device to operate. One of the principal requirements is that under any circumstances the duration of the test current cannot exceed 2 seconds, this makes it possible to test most time delay devices which have been installed to ensure discrimination between RCDs of different operating currents. In addition most RCD test instruments have the facility to carry out the test in either the positive or negative half cycles of the supply. Each of the tests should be carried out in both half cycles and the highest value obtained should be recorded.

> *Remember*
> The test instrument will need to have leads suitable to carry out the tests on circuits supplying both socket outlets and fixed equipment. We shall need leads with both a moulded plug and split leads, similar to those used for the earth fault loop impedance tester.

There are three basic tests which should be carried out. In this instance we are going to consider the testing of a device rated at 30 mA.

Half rating test

The half rating test is intended to establish whether the device is likely to suffer from nuisance tripping under normal operation. The RCD test instrument is set to half the rated tripping current for the device under test, in our example this would be 15 mA, and the instrument is connected as shown in Figure 5.48.

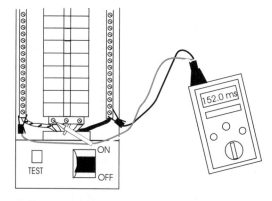

152.0 ms

ON

TEST

OFF

Figure 5.48

The test is then carried out and the result noted. For this test the instrument should apply the test current for the full 2 seconds without the RCD operating. If the device should trip it is possible that there could be nuisance tripping and the manufacturer of the device should be consulted to establish the operating parameters of the device and whether a replacement should be fitted.

Trip test at 100%

The instrument is then set to the full operating current of the device, in this case 30 mA, and the test applied. This time the device should operate and the time taken displayed on the instrument, this should be noted and the device reset. The instrument should then be set to test on the other half cycle and the test repeated and again the result should be noted. The highest of the two results is the one which should be recorded on the Schedule of Test Results.

We need to check the recorded values against the maximum operating times for the type of device installed. As the criteria for acceptable operating times will depend on the product standard we shall need to refer to the maximum for the particular standard. The table shown in Table 5.2 gives the maximum values for the product type when tested at the full rated current for some of the most common RCDs.

Table 5.2

RCDs to BS 4293 RCD protected sockets to BS 7288	Less than 200 ms
RCDs to BS 4293 (with a time delay incorporated)	Between: 200 ms + 50% of the time delay and 200 ms + 100% of the time delay
RCDs to BS EN 61008	Less than 300 ms

Trip time at 150 mA

The third test to be carried out is applicable to RCDs used to provide protection against direct contact and rated at no more than 30 mA. For these devices we must carry out a test at five times the rated current, so in the case of our 30 mA device this will be at 150 mA. When this test is applied the device should operate within a time of 40 ms, that is 0.04 seconds.

Once the above tests have been completed the device should be reset and the test button operated to ensure the device operates.

Remember

The RCD test button is the quarterly check that should be undertaken by the user of the installation and simply checks the device operation. It does not confirm satisfactory disconnection times for protection against electric shock.

Try this

Sketch the diagram for the testing of an earth electrode using a proprietary earth electrode test instrument.

Functional tests

We are expected to carry out functional tests of equipment such as switches, interlocks, drives and controls to ensure they operate correctly. This will include checking the operation, mounting, accessibility and adjustment of equipment and devices. There is no formal record of each item required however there is a need to indicate that the functional testing of equipment has been undertaken.

Having completed the testing we need to ensure that the results are accurately recorded on the appropriate schedule. The details required are self explanatory if the headings are followed, and further guidance is provided later in this chapter. It is always advisable to have a copy of the results schedule available for the recording of the results on site. These may then be transferred onto the schedule of test results which forms part of the certificate or report to be issued to the client. Besides providing a ready reference for the recording of test results it also makes a useful aide-mémoire to ensure that all the tests are completed before we leave site.

When we have completed the inspection and testing of the electrical installation and determined that it is safe to put into service we need to record this fact. To do this we need to complete a form of certification or reporting, depending on the nature of the work we have undertaken. We shall need to issue a certificate for a new installation, and we may also be asked to consider an existing installation for

- an alteration or addition
- change of use
- a particular licensing application
- compliance with the requirements of BS 7671 and whether it is safe for continued use

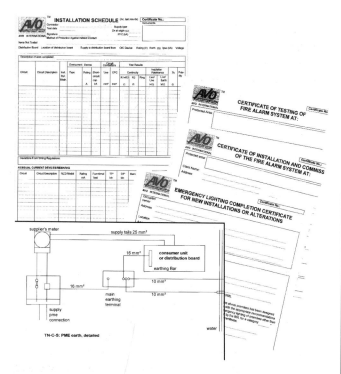

Figure 5.49   Records must be kept

Figure 5.50

When any of these activities are undertaken we must issue some form of certification to show what we have inspected and tested. Equally important is to record those things which we did not inspect or test. These documents should be passed to the client and copies retained on file by the electrical contractor.

These documents show

- the compliance of a new installation, addition or alteration with BS 7671
- the tests undertaken to confirm compliance
- the extent of the work to which the documentation applies
- any exclusions or limitations that were made
- the date of the work
- the persons responsible for the work
- the standard of an existing installation
- the extent and limitations placed on the assessment of an existing installation

In addition to this they also provide evidence that the requirements of the Electricity at Work Regulations have been met.

In the unfortunate event of an incident occurring, on an installation on which you have worked, the certification and reporting documents are the only evidence you have of the items listed above. Without the documentation there is no evidence that any of the statutory obligations were met or that any safety checks were taken, even the extent of the work undertaken cannot be confirmed. If additional work has been carried out since your installation was completed there is no way for you to prove this was not your responsibility.

The requirement to maintain records applies to all aspects of the installation work from design to commissioning and placing into service. Before we consider the final documentation required we need to review the requirements for the records collected during the design and construction phases of the installation. In the first instance we shall consider the requirements for a new installation. This requirement will also apply to additions and alterations to an existing installation as the requirements are the same for all electrical installation work.

Design data

During the course of the design of the electrical installation we need to determine the requirements of the installation for its intended use. This will include the client's requirements, the loading of the equipment to be installed, the maximum load, the characteristics of the supply and any external influences which may affect the electrical installation. Upon this background of information we can progress the design and select wiring systems, control and protection devices, cable types and sizes, working to the final design for the installation. We need to maintain a record of all the information that we use to design the installation.

Any information used for changes to the design, and any calculations supporting the final solution, should also be recorded. Manufacturers' details confirming power requirements, operational criteria and the like should be retained and used for the design. These details should also be included with the documentation which is issued to the client on completion of the work.

Certification

On completion of the installation we are required to issue, to the person ordering the work, our client, a certificate appropriate to the work we have carried out. The requirements for certification are defined in Part 7 of BS 7671 and Guidance Note 3. Any certification issued should contain at least the information required in the standard forms set out in BS 7671 and be accompanied by a schedule of test results, detailing the design and test details of each circuit. Electrical contractors are free to develop their own forms of certification or buy ready printed forms. It is always advisable to check that the forms of certification to be used provide all the information required by BS 7671 as a minimum.

We shall now look at the various forms of certification, their application and completion, beginning with the:

Electrical Installation Certificate

This certificate is used to certify new electrical installation work. It may be used to certify a complete installation or additions and alterations to an existing installation. It should be issued by the contractor responsible for the construction of the electrical installation and a separate certificate should be issued for each distinct installation. A certificate must be issued to the person who ordered the work irrespective of whether they, or their client, have requested a certificate. The guidance for "recipients of certificates" should be brought to their attention. This requires the certificate, or a full copy of it, to be passed immediately to the user.

In this chapter we shall be considering the minimum content required by BS 7671. It would be advantageous to have a copy of both BS 7671 and IEE Guidance Note 3 available for reference.

The certificate should detail the following;

Details of the client and the installation

The client is the person who ordered the work, your client, who may be the main contractor, the architect or principal electrical contractor. The client will not always be the owner or indeed the end user of the installation. The installation details should contain sufficient information to allow the installation to be clearly identified. The information should also state whether the installation is new, an alteration or addition and quite often more than one of these categories is applicable. New circuits may be installed along with additional and alterations to existing circuits when buildings are modernised or extended.

The extent of the installation which is covered by the certificate should be clearly identified. This information identifies the extent of the work for which you are responsible and should be completed in every case with a clear description. It is acceptable to identify either the extent of the work undertaken or any exclusions from the certificate whichever is the lesser. For example a complete rewire of a commercial property with the exception of the car park lighting would be better described by identifying "the electrical installation at *the address*, and excluding the electrical installation and the control equipment for the car park lighting", in the extent section of the form.

Design, construction and inspection and testing

The Electrical Installation Certificate should be issued by the person responsible for the construction of the installation. However, the same person may not be responsible for all aspects of the design, construction and inspection and testing of the installation. The certificate contains a declaration that all these aspects of the electrical installation are in accordance with the requirements of BS 7671. There are two options available:

1. Where all aspects of the design, construction and inspection and testing are carried out by the same person, there is the provision of a single declaration box for all three aspects and therefore requires just a single signature.

2. Where the responsibility for the design, construction and inspection and testing of the installation is not by the same person then each part must be declared separately. A separate section is included for each aspect and should be signed accordingly. The design of the installation may not be carried out by one person, for example where a consultant provides some aspect of the design and distribution circuits, and the final circuit design is carried out by the contractor. In such circumstances there is provision for each individual to sign for their design, the extent of the design for which each is responsible is not detailed on the certificate. The design data produced by each party serves to provide this information, should it be required at a later date.

Supply characteristics and earthing arrangements

In this section of the certificate we detail the information relevant to the supply for the installation, most of this detail we had to establish for the design of the installation. Here we confirm the details as they are on site. This includes;

- the type of system, TN-S, TN-C-S or TT being the most common types
- number and type of live conductors. Remember this includes neutral conductors but not earthing conductors
- nature of the supply. This includes the nominal supply voltage, the frequency, external earth fault loop impedance and the prospective fault current.

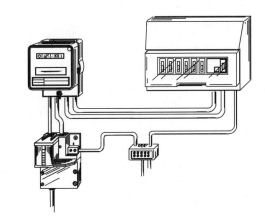

Figure 5.51   Main intake

Particulars of the installation at the origin

Here we provide the information particular to the main intake of the installation. We need to include information on the means of earthing, details of the earth electrode where one is fitted, the earthing and main bonding conductors, the maximum demand and the main switch or circuit breaker. The details should include items such as current ratings, BS type, location and csa.

Comments on the existing installation

When the certificate is used to certify alterations and additions to an existing installation then it is important to record any defects in the existing installation which do not affect the safety of the installation. Defects which would result in a reduced level of safety for the new installation must be rectified before the new installation is placed into service.

When the installation is new, or there are no defects found in the existing installation, then this section should be completed as "none observed".

Next inspection

Once the installation is completed, the electrical installation certificate should provide information on the date at which the next inspection and test is due. IEE Guidance Note 3 gives some guidance on the maximum period between inspection and test for installations in a variety of types of premises. These values are the maximum periods and shorter periods may be appropriate dependent on the use, location and environmental influences which affect the installation. If a shorter period is recommended then we must be able to justify why this is required.

Schedules

Included with the certificate should be the schedules of items inspected and tested and the schedule of test results.

Items inspected and tested

The schedule items relate to the items which need to be inspected and tested in accordance with BS 7671. However, not all the items required by BS 7671 occur on every installation. Those which do not apply should be recorded as not applicable (N/A). Common examples of these would be Obstacles and Placing out of reach. These are methods of providing direct contact protection and are only used in special circumstances. It is important to record all the items which have been inspected and tested and to ensure that these cover all the requirements of BS 7671.

AVO™ AVO INTERNATIONAL

INSPECTION SCHEDULE

Certificate No.:

Items Inspected

1. Connection of Conductors
2. Identification of Conductors
3. Routing of cables in safe zones or protected against mechanical damage
4. Selection of conductors for current/voltage drop
5. Connection of single pole devices for protection or switching in phase conductors only
6. Correct connection of socket-outlets and lamp holders
7. Presence of fire barriers and protection against thermal effects
8. Method of protection against electric shock
   a) Protection against both direct and indirect contact
      - SELV (note 1)
      - Limitation of discharge of energy
   b) Protection against direct contact (note 2)
      - Insulation of live parts
      - Barrier or enclosure
      - Obstacles (note 3)
      - Placing out of reach (note 4)
      - PELV
   c) Protection against indirect contact
      (i) EEBAD including:
      - Presence of earthing conductors
      - Presence of protective conductors
      - Presence of main equipotential bonding conductors
      - Presence of supplementary equipotential bonding conductors

      (ii) Use of Class II equipment or equivalent insulation (note 5)

      (iii) Non-conducting location (note 6): Absense of protective conductors

      (iv) Earth-free local equipotential bonding (note 7): Presence of earth-free equipotential bonding conductors

      (v) Electrical separation (note 8)
9. Prevention of mutual detrimental influence
   - Proximity of non-electrical services and influences
   - Separation of Band 1 and Band 2 circuits or Band 2 insulation used
   - Separation of fire alarm and emergency lighting cables
10. Presence of appropriate devices for isolating and switching correctly located
11. Presence of undervoltage protective devices where appropriate
12. Choice and setting of protective and monitoring devices (for protection against indirect contact and/or overcurrent)
    - Residual current devices
    - Overcurrent devices
13. Labelling of protective devices, switches and terminals
14. Selection of equipment and protective measures appropriate to external influences
15. Adequacy of access to switchgear and other equipment
16. Presence of danger notices and other warnings
17. Presence of diagrams, instructions and necessary information
18. Erection methods
19. Requirements of special locations

Inspected by ....

Date(s) Inspected ....

Notes

1. SELV An extra-low voltage system which is electronically separate from earth and from other systems. The particular requirements of the Regulations must be checked (see Regulation 411-02)
2. Method of protection against direct contact - will include measurement of distances where appropriate

5. Use of Class II equipment - infrequently adopted and only when the installation is to be supervised (see Regulation 413-03)
6. Non-conducting locations - not applicable in domestic premises and requiring special precautions (see Regulation 413-04)

Figure 5.52 *Inspection schedule form*

Schedule of test results

AVO AVO INTERNATIONAL

INSTALLATION SCHEDULE (inc. test results)

Certificate No.:

Instruments:

Contractor ....

Test date ....

Signature ....

Supply type ....

Ze at origin (Ω) ....

PFC (kA) ....

Method of Protection Against Indirect Contact ....

Items Not Tested ....

Distribution Board	Location of distribution board	Supply to distribution board from	O/C Device	Rating (A)	Earth (Ω)	Ipsc (kA)	V

Description of work completed ....

Circuit	Circuit Description	Inst. Ref. Meth.	Overcurrent Device			Circuit Conductors		Test Results					
			Type	Rating	Short-circuit cap.	Live	CPC	Continuity			Insulation Resistance		Zs
								R1+R2	R2	Ring	Live/Live	Live/Earth	
				A	kA	mm²	mm²	Ω	Ω		MΩ	MΩ	Ω

Deviations From Wiring Regulations:

RESIDUAL CURRENT DEVICES/REMARKS

Circuit	Circuit Description	RCD Model	Rating mA	Functional Test	TT* Idn	OT* Idn	Remarks

Figure 5.53 *Schedule of test results*

The schedule of test results is where we record the information regarding each individual circuit, detailing the design criteria and test results. Larger installations may well need to have a number of schedules of test results produced, and providing each schedule clearly identifies the part of the installation to which it applies, this is an acceptable way of providing this information.

The first information we need to provide is that for the distribution board and we need to include where it is located and what identification reference it has been given. We also need to identify where it is supplied from, the nominal voltage, number of phases and the type and rating of the protective device for the distribution circuit.

Where the distribution board is remote from the origin of the installation we also need to record the values of earth fault loop impedance and prospective fault current measured at the distribution board.

Where the distribution board is located at the main intake position this information will be as entered on the details of the installation at the origin and so will not need to be entered again here. Finally we need to record details of the test instruments used, including the make and serial number for reference purposes.

It is generally easier to consider the test schedule in two parts, the first related to the circuit design details and the second to the test results. As an example, let's consider a single circuit supplied from a distribution board. The design aspects of the circuit are those which relate to the physical construction of the circuit including details such as the:
- circuit number and phase
- circuit description, which should be as descriptive as possible in order to distinguish one circuit from any other (lights, lights, sockets, sockets and so on are not suitable descriptions)
- type, rating and short circuit rating of the protective device, for example BS EN 60898, type "B", 16A, 9 kA
- reference method for the installation of the circuit, as detailed in BS 7671
- type of wiring, for example PVC, PVC
- size of the live and protective conductors for the circuit.

This information should have been established at the design stage of the installation and would be confirmed on site.

The "second" part of the schedule of test results contains the results of the actual tests undertaken on the installation. We shall need to use these results to ensure that the installation complies with the design and the requirements of BS 7671.

We have covered the testing requirements earlier in this chapter and so we shall not be considering them again here. However we shall consider the requirements related to the recording of the results.

The test results which need to be recorded are normally laid out in the sequence in which the tests are carried out. The sequence is detailed in BS 7671 and the majority of test schedules include those tests which are carried out for all installations. Any specialist installations, or parts of installations, are usually separately certified by the specialist responsible for carrying out the tests.

So we need to record:

Continuity

We would normally record this as $R_1 + R_2$, this being the most effective method of measuring the continuity of circuit protective conductors. Where the wander lead method of measurement is used we would need to record the R_2 value and clearly indicate that this is the value recorded.

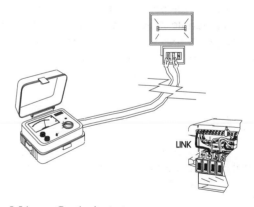

Figure 5.54   Continuity test

We should also record the ring circuit continuity values for any ring circuits. Where this section is not applicable as the circuit is not a ring circuit then we should enter N/A. Some schedules contain columns for recording all three values, R_1, R_n and R_2 and as these results need to be taken on site, to confirm ring circuit continuity, it is advisable to record the values on the schedule.

Insulation resistance

We need to record the insulation resistance and ideally the full range of test values should be recorded for each circuit, that is phase to phase, phase to neutral, phase to earth and neutral to earth. Once again where the p-p value is not applicable, on single phase circuits, we should enter N/A.

Polarity

We need to indicate that the polarity has been tested and is correct. The circuit should not be put into use if the polarity is incorrect. This is normally indicated by the means of a tick against each circuit once the polarity has been both tested and checked to be correct.

The above tests are carried out **BEFORE** the circuit is energised to allow the remainder of the tests to be completed in the knowledge that the installation is safe to energise. This is important for the safety of the people carrying out the testing and anyone in the vicinity at the time the tests are carried out.

> *Remember*
> The $R_1 + R_2$ test for continuity provides a positive indication for the polarity of the circuit. Once the circuit has been energised we need to check that the polarity is correct, this is normally done when the live tests are carried out. The appropriate boxes on the schedule of test results can only be ticked when the polarity has been both tested before **and** checked after the circuit is energised.

Earth loop impedance

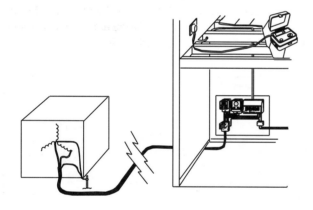

Figure 5.55   Earth fault loop impedance test

We need to test and record the maximum earth loop impedance for the circuit, and this is generally the value measured at the furthest point from the supply on each circuit. Measuring Z_s at each outlet will determine the maximum value for each circuit and it is this maximum value which needs to be recorded.

Operation of RCDs

There are two values which may need to be recorded here, the first being the operating time for the RCD at the rated current, so the disconnection time measured, for say a 30 mA RCD at 30 mA, would be recorded. The second is the operating time for an RCD of 30 mA or less which is installed to provide supplementary protection against direct contact. 30mA RCDs

installed to provide protection for socket outlets likely to supply portable equipment for use outdoors would be a typical application. In these cases the operating time of the RCD when tested at $5 \times$ the rated tripping current needs to be recorded.

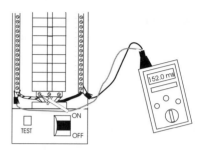

Figure 5.56   RCD test

> *Remember*
> Many RCD test instruments have the facility to carry out the test in both the positive and negative half cycles of the supply. The tests should be carried out in both half cycles and the higher of the two values obtained should be recorded. This represents the most onerous condition for the circuit. The results are recorded in milliseconds.

The test results obtained should be verified against the requirements of BS 7671 and the original design parameters in order to ensure that the installation complies with those requirements and that it is safe to be put into service.

Having considered the Electrical Installation Certificate we will now look at the Minor Electrical Installation Works Certificate.

> *Try this*
> Using the Inspection Schedule in Guidance Note 3 as reference, identify the items which would not normally be applicable to a domestic installation.

Minor Electrical Installation Works Certificate

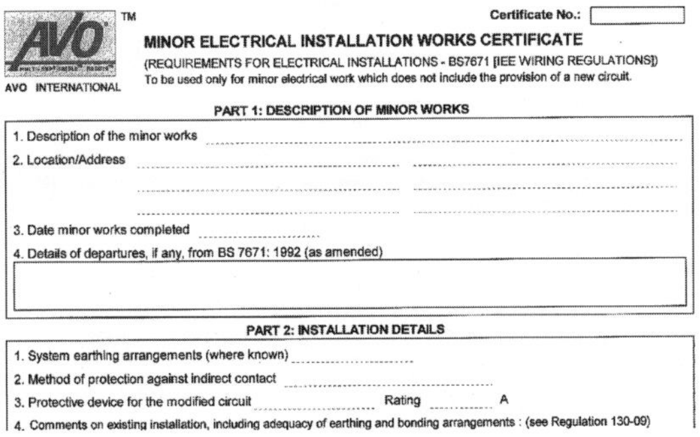
AVO™ AVO INTERNATIONAL

Certificate No.:

MINOR ELECTRICAL INSTALLATION WORKS CERTIFICATE
(REQUIREMENTS FOR ELECTRICAL INSTALLATIONS - BS7671 [IEE WIRING REGULATIONS])
To be used only for minor electrical work which does not include the provision of a new circuit.

PART 1: DESCRIPTION OF MINOR WORKS
1. Description of the minor works
2. Location/Address
3. Date minor works completed
4. Details of departures, if any, from BS 7671: 1992 (as amended)

PART 2: INSTALLATION DETAILS
1. System earthing arrangements (where known)
2. Method of protection against indirect contact
3. Protective device for the modified circuit .... Rating .... A
4. Comments on existing installation, including adequacy of earthing and bonding arrangements : (see Regulation 130-09)

Figure 5.57 Minor works form

This certificate is intended to certify modifications to a single circuit only, where a new protective device is NOT installed, so it must not be used to certify a new circuit. The Minor Electrical Installation Works Certificate has some similarities with the Electrical Installation Certificate, although by the nature of the certificate the content is somewhat reduced, occupying a single page.

Part one of the Minor Electrical Installation Works Certificate is where we record the details of the location of the work and the extent of the work undertaken. We also need to record the date the work was completed and any departures from BS 7671.

Part two of the certificate is used to record the details of installation which includes the type of earthing system, the method of protection against indirect contact, the type and rating of the existing protective device for the modified circuit and any comments on the existing installation.

Part three of the certificate is used to record the test results for the circuit. These are those tests essential to ensure the circuit is safe to put back into service. These include earth continuity, insulation resistance, earth fault loop impedance, polarity and RCD operation where one is fitted.

Part four of the certificate is the declaration where the person responsible for the works declares that the work does not impair the safety of the existing installation. The signatory also declares that the work carried out has been designed, constructed, inspected and tested in accordance with BS 7671 and that the work complies with the requirements of that standard.

The Minor Electrical Installation Works Certificate is designed for use under particular conditions and the extent of the information provided on the certificate reflects that. Where minor works are undertaken, the original existing installation should have an Electrical Installation Certificate and the Minor Electrical Installation Works Certificate is supplementary to that certificate.

Under the specific circumstances covering the issue of the Minor Electrical Installation Works Certificate we are verifying the suitability and safety of the modified circuit to be put into service. It is important to ensure that the existing installation

- is suitable for the intended work,
- will not have its safety compromised on completion of the proposed work

and

- the circuit we modify is compliant with the requirements of BS 7671.

This sometimes causes confusion as to the extent of the work necessary to ensure compliance with BS 7671.

As an example let us consider the standard domestic installation, where we are to modify an existing power circuit supplying the ground floor socket outlets. Before we begin the work, we need to establish the means of protection against indirect contact which is generally EEBAD (Earth Equipotential Bonding and Automatic Disconnection). Assuming this to be the case, we must ascertain that the existing installation has the necessary main equipotential bonding in place to provide that protection. If this is not in place or is not adequate then the bonding should be brought up to the required standard before the circuit is modified and put back into service. Similarly we need to consider whether any of the sockets on that circuit are likely to be used to supply portable electrical equipment outdoors. If so the circuit should be protected by an RCD rated at no more than 30mA and if one is not already fitted then this would need to be installed before the circuit is modified and put into service.

Remember

The circuit modified and certified using a Minor Electrical Installation Works Certificate must meet the requirements of BS 7671 and not impair the safety of the existing installation.

Those parts of the existing installation which affect the compliance and safety of the modified circuit must also meet the requirements of BS 7671.

Other aspects of the existing installation which are observed that are not in accordance with BS 7671 and do not affect the compliance of the modified circuit should be brought to the client's attention and recorded on the Minor Electrical Installation Works Certificate.

You have an obligation under the Electricity at Work Regulations to ensure that the duty holder for the installation is informed, without delay, of any dangerous aspects of the installation which could affect the safety of the users of the installation.

Figure 5.58   The periodic report form

The Periodic Inspection Report

This is different from the previous Certificates because it is not issued for electrical installation work. As the name implies, it is used to report on the condition of an existing electrical installation, one that has been placed into service. The report is used to record the compliance of an installation, which has been in use, with the current requirements of BS 7671.

Remember

A Periodic Inspection Report records the compliance of an existing installation with the CURRENT requirements of BS 7671, irrespective of the age of the installation or what regulations were effective at the time the installation was completed.

So why does an installation require a periodic report? Every electrical installation deteriorates with age and use. It is important to ensure that the installation continues to be safe for use and does not put users of the installation at risk. The frequency of the inspection is dependent upon a number of factors, such as the type of building and its use, the age of the installation, and the environmental conditions. IEE Guidance Note 3 provides some guidance on the maximum period to the first inspection and test of a new installation based upon these criteria. The frequencies between inspections are determined considering these factors, and the condition of the installation following a period of use.

The purpose of the periodic inspection is to establish, as far as is reasonably practicable, any factors which could impair the safety of the electrical installation and report on them. Chapter 73 of BS 7671 outlines the requirements for periodic inspection and testing of electrical installations.

So having established the need for periodic inspection and testing and the need to report on the findings, what of the format used for the report? Appendix 6 of BS 7671 and Part 5 of IEE Guidance Note 3 give standard forms for the periodic inspection and testing of an electrical installation. In addition to the standard forms we need to produce a schedule of items inspected and tested and a schedule of test results. The schedules and their compilation are generally identical to those for the Electrical Installation Certificate and so we shall not be considering the schedules again.

The first part of the form provides details of the client and their address, which may be different from that of the installation to be inspected and tested, and the reason for which the report is required. It is important to record the purpose for which the report is required. This may be for a for a restaurant, hotel, public building or a particular use such as a public entertainment licence. It may be required as part of a house sale survey or as a part of the duty holder's insurance requirements or legal obligation.

The next section of the form is where we record the details of the installation. This requires information such as the occupier and the address of the installation.

Remember

The client may not be the occupier and they may not reside at the same address as the installation to which the report refers.

We must also record some information relative to the installation itself which includes the

- type of premises
- estimated age of the electrical installation
- evidence of any alterations or additions to the installation
- date the last inspection and report was carried out
  and
- whether there are records available

This information provides details of the existing electrical installation upon which the periodic inspection and report is carried out and should always be filled in as accurately as possible.

The next box is where we detail the extent and limitations of the inspection and report and is one of the most important parts of the report. It is here that we have the opportunity to detail what was, or was not, included in the inspection and any limitations placed upon us whilst we were carrying out the inspection.

These items should be agreed with the client before work is started so that both parties are aware of the extent of the work to be undertaken. The requirements of any third party, such as an insurance company or licensing officer must, of course, be taken into account. The overall extent of the work is however the subject of negotiation between the electrical contractor and the client. The contractor provides the technical advice as to the electrical and any additional third party requirements. Once these are agreed the proposed extent of the work and any foreseen limitations can be noted.

The true extent of the work and any limitations can only be recorded on the report once the work has been completed. If, for example, at the time of the inspection the user of the installation restricts your access, or parts of the installation cannot be isolated, then these items will need to be recorded. Then the recipient of the report will be accurately informed of the extent of the inspection and to what the report actually refers.

The next two sections can only be completed once the inspection and testing has been carried out and the condition of the installation established. The date of the next inspection is, as we found earlier, dependent upon the condition of the installation, the environment, the type and use of the premises. The actual period to the next inspection is therefore based upon your findings and the period cannot be determined until the report has been completed. The second section is the declaration which, once again, cannot be completed until the inspection and testing has been carried out and the condition of the installation established.

The next section of the report deals with the aspects of the installation which need to be measured or ascertained and are similar to those at the origin of the installation recorded on the Electrical Installation Certificate and so we shall not be covering these again here. However there is one section which requires some discussion and clarification and that is the Observations and Recommendations section.

Observations and recommendations

Figure 5.59   *Observations and recommendations section from page two of the periodic report form*

This is the section of the report where you make your observations on the installation and recommendations as to the action required. There are two important factors to bear in mind here.

Try this

a)   State the purpose of a Periodic Inspection and Report.

b)   What other factors are there which need to be considered when a periodic report is to be carried out?

Observations

These should be observations based upon your findings and relate to those parts of the installation which do not meet the requirements of BS 7671. We are not required to record the regulation numbers as the report should be understood by a layperson. However we should be able to support each observation by a regulation with which the installation does not comply. Custom and practice should not be included in the observations, just because a particular aspect of the installation is not done as you would normally do it does not automatically mean it does not comply with the requirements of BS 7671.

Recommendations

The recommendations are not instructions on how to correct the departures or information on what needs to be done. There is a standard code for recommendations and these are in the form of numbers 1,2,3 & 4 to indicate the action that needs to be taken.

In simple terms the recommendation codes are:

Code 1:
> Requires urgent attention: informs the recipient that potential danger exists and that this item requires immediate action.

Code 2:
> Requires improvement: informs the recipient that a deficiency exists in the installation which does not currently represent a danger but which requires improvement.

Code 3:
> Requires further investigation: informs the recipient that, within the agreed extent and limitation of the report the inspector was unable to come to a conclusion on this aspect of the installation. It is important to remember that the purpose of the inspection and report is not to carry out fault finding. If a circuit is found to be below the requirements for insulation resistance testing, for example, this should be recorded but the precise cause of the failing would not normally be found.

Code 4:
> Does not comply with BS 7671 as amended: indicates that there is a departure from the requirements of BS 7671 but the users of the installation are not in any danger. A typical example would be where green only sleeving is used on the cpcs of a circuit whilst BS 7671 requires these to be coloured green and yellow.

It is important that the Observations and Recommendations are recorded in this way to provide standard information and recommendations to those receiving the reports.

Remember
There may be several alternative methods of achieving the necessary improvement required. The purpose of the report is to identify the need for improvement and the urgency with which this action needs to be undertaken. The precise remedy is a matter for discussion between the person responsible for the installation and the electrical contractor engaged to carry out the remedial work. Both will be using the information provided on the report to determine the most suitable and cost effective action to be taken to achieve a compliant installation.

The other sections we need to consider are the summary of the inspection and the schedule record.

The summary of the inspection is where we describe the overall condition of the installation. The comments made here should adequately describe the condition of the installation and help to substantiate your overall recommendations. If, for example, some aspects of the installation, whilst compliant at the time of the inspection, are showing signs of deterioration then this should be recorded here. It would be reasonable in such circumstances to recommend a shorter period to the next inspection as a result. Simple statements such as "poor condition" or "needs rewiring" do not adequately summarise the condition of the installation or provide sufficient information to substantiate your statement.

Finally we need to record the number of pages that accompany the report. These will include the schedules of items inspected and tested, the schedules of test results and any additional pages used to record the observations and recommendations or summary of the installation.

Remember
Always consider your observations, recommendations and summary in such a manner that, if you were to receive the report, you could provide a reasonable quotation for the remedial work, having never seen the installation. The information provided should therefore be sufficient to allow you to make a reasonable assessment of the actual condition of the installation.

PROJECT

Now we have completed Chapter 5 of the module we should consider parts 4a, 4b and 4c of the project. In this we shall need to produce a distribution diagram for the installation. This will cover the mains and sub-mains cables and will be a schematic drawing. We shall also be considering the requirements for certain aspects of testing so reference to Guidance Note 3 may be advantageous.

6

Estimating

In this chapter we shall look at the requirements for estimating and tendering for work. Many large companies employ estimating engineers whose sole function is to estimate for new work and advise for quotations. In small companies it is often the supervisor responsible for the work who also prepares estimates and tenders.

Figure 6.1

On completion of this chapter you should be able to:

- ◆ assess quantities of material, labour and plant from drawings, specifications and measurement
- ◆ state problems associated with delivery, access, handling, storage and siting of materials, and suggest methods of overcoming such problems
- ◆ carry out a take off of quantities from given data
- ◆ state considerations to be made during estimates
- ◆ state additional management functions on estimate to produce tender
- ◆ state the purpose of Bills of Quantities

Before we go any further we should consider the difference between an estimate, a tender and a quotation. All three are related to providing the client with a cost for carrying out work, but each has a distinct difference in application and implication.

Estimate:

This is the basic direct cost of a project in terms of materials and labour. This is the cost that a company believes it will incur, in order to carry out the work.

Tender:

This is the result of the subsequent management actions on an estimate. This takes into account all the other relevant factors in order to determine the price for which the company is prepared to carry out the work.

These factors will include:
- The size of the job and does it carry any prestige?
- The relationship with the client. For example, is this a new client? are works already being carried out for the same client elsewhere?
- The competition from other contractors and their current trading position within the industry.
- Are there additional works which may follow on from this job?
- The level of profit or loss that the company is prepared to accept for the work.

Whilst this is not an exhaustive list, it does indicate the type of considerations that have to be made to produce a Tender figure.

Quotation:

For small works the overall tender figure is often referred to as a Quotation. This is the sum that the contractor agrees will be the actual charge for the work he has been asked to do. It is important that the extent of the work included is quite clear and unambiguous.

Figure 6.2

Let's suppose we are asked to carry out a small electrical installation, such as an electric shower. The cost advice we give to the client may be as one of two alternatives:

The estimate:

which is an approximation of what the contractor believes the work will cost. This is not usually the final cost, but a best guess as to what the cost will be. It does not usually allow for unforeseen problems. Should any arise the contractor will seek additional payment to cover the costs. This use of the term estimate is often applied to smaller work where no written contract exists.

The quotation:

which is a full figure for carrying out the work. Any problems that are encountered are deemed to be included. The contractor can only request additional monies where the client requests a change to the original work. If this involves the contractor in additional cost then this can be claimed. Equally if the change results in a saving the client should receive a reduction in the overall cost of the work, in line with the saving.

On larger contracts the work is sent out for Tender which is essentially the same as a Quotation. The company winning the contract agrees to carry out all the works as detailed for a fixed sum. It is this fixed sum factor which can make the estimating and tendering for work so important. If a significant item of equipment, such as the main switch gear, is omitted from the estimate, it is still deemed to have been included in the tender.

For the most part we shall be considering estimating for new works and simply refer to factors which will form part of the "commercial decision" process.

The receipt of invitation to tender

When work is put out to tender, it is usually in the form of documents and drawings which together detail what is required. Depending on the size of the works, the documentation can vary from a brief list written on a cigarette packet to three or four volumes. The drawings could be anything from a builder's free-hand sketch through to fifty or more "A0" size drawings. For our purposes we shall consider that some documents and drawings have been received.

When an invitation to tender arrives we must note all the documents and drawings which have been received. This will involve recording all the identification numbers, any revisions and document identification references. These should then be checked against the list of information referred to in the invitation to tender to make sure everything we should have is there.

The estimate

The first thing that we must do is read all the documentation. Whilst doing so we must determine whether there are any particular or unusual requirements. Overlooking a particular requirement at this stage could cause problems later on and affect the tender price. It is therefore important that we pay close attention to detail at this stage.

Once we have carefully examined the documents and made a note of any particular requirements our next job is to examine the drawings. We should note down the details of any major items of equipment such as generators or switch gear as we go through. We need to send these details to the manufacturer(s) or supplier(s) in order to obtain a quotation(s) for supplying the equipment. This will be included in our tender price.

We need to take note of any particular manufacturer who may be specified within the tender documents. We must obtain a quote from that particular manufacturer. However, we can also ask for quotes from other companies to find out if the equipment can be purchased at a lower price. Should we find that we can we should inform the client of this when our tender price is submitted.

A list price for standard items such as socket outlets and switches is generally available. We may also be able to negotiate a better discount from suppliers or wholesalers, if items are required in large numbers.

Our next task is a "take off" of the materials required from the drawings. The quantities are put into a table under their various headings, to enable us to calculate the total quantity of each item required. There are a number of ways of doing this and most companies devise their own form for this operation. Figure 6.3 shows a typical layout of such a form.

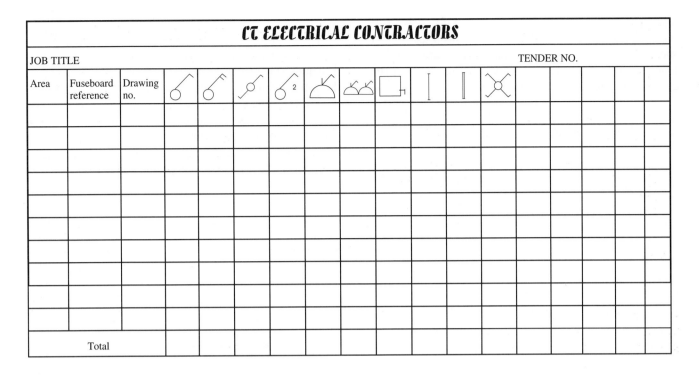

Figure 6.3   Basic Take Off form

The horizontal headings across the top contain the standard symbols for items of equipment and accessories. There are always blank columns or sheets to allow the estimator to put down any non-standard requirements. The first vertical column is to record the area for which the details are taken. We then record any fuseboard references and then the drawing number.

Once the details are recorded we can add up the vertical columns to obtain the total number of each item required. If we wish to check the quantities for any particular area then these are contained in the horizontal row for that particular drawing, as in Figure 6.4. This is useful should the contract be won, as we can arrange deliveries on a floor by floor basis using the detail in the tender.

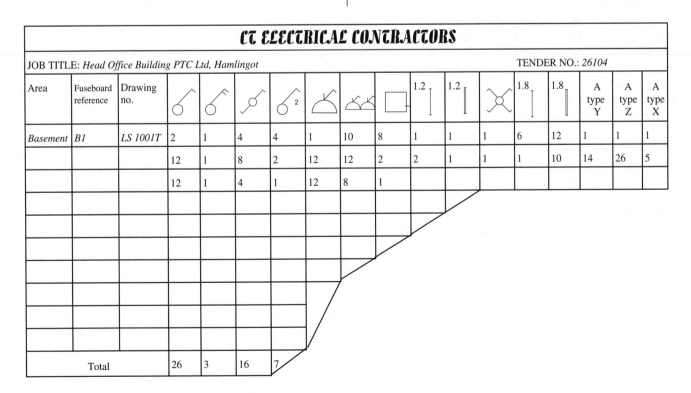

Figure 6.4   Take Off form partly completed

Once we have completed our detailed list of materials we can obtain our material cost. The summary totals of our take off sheets are then compiled on a summary listing form to obtain a total cost. This may be carried out using a form similar to that shown in Figure 6.5.

Remember

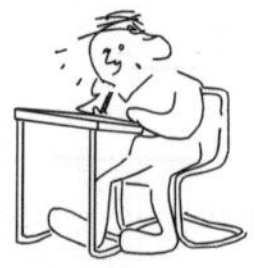

As we work through each drawing it is a good idea to strike through each item with a pen or pencil as it is counted. Should you be called away write the total so far against the last item counted so that you can resume your count later.

COST ESTIMATE SHEET No. of sheets

TENDER NO.

JOB TITLE

TAKE OFF SHEET NUMBERS

Symbol	Specification	No. off	Materials £						Labour manhours					
			Rate	Total					Rate	Total				

Figure 6.5 *Cost estimate sheet*

Bill of Quantities

At this point we should consider a tender where we are supplied with a Bill of Quantities. Where Bills of Quantity are provided at tender the drawings provided are generally layout diagrams, often to a small scale. We are not required to carry out a take off as the client's representative, usually a quantity surveyor, has carried out this function. The client then provides the list of materials required.

We then have to obtain our best price for the materials and add to it the labour element. This is then the basic estimate and we would then progress as we would for a take off tender. The Bill of Quantities tender document makes provision for the recording of material and labour costs involved and a typical Bill of Quantities tender record is shown in Figure 6.6.

The next step is usually to assess the labour needed to install the material on a basis of hours per item /metre etc. We must not forget to allow for incidental labour required to carry out works not directly associated with the material count. The quantities are then recorded on a summary sheet as in Figure 6.5.

At the end of this process we will have estimate figures for both labour and material. If we have been asked to provide a fixed price for the works any increase in these two figures, between the start and finish of the work, must be allowed for and included in the tender price.

Remember
The actual task of calculating the cost of labour for a contract can be very complex. There is a standard time for nearly every operation an operative will carry out. An estimator uses these and his experience to arrive at a figure which is right for the specific conditions.

During the process of estimating we must use our experience of installation work to assess any problems that may be encountered. These are quite varied but are generally associated with the type of building construction and the wiring system to be installed. It would be common practice for a warehouse installation to be done in steel conduit and trunking.

PROJECT NAME			NO.	
	Quantity	Unit	Each	Total
Fire detection and alarm		nr		
Electrical services measurement				
Conduit and cable trunking				
Galvanised steel tray including all fittings, fixings and supports				
Cable trays				
150 mm wide	348	m		
HV/LV cables and wiring				
Fixed to surface of tray				
Pyrotenax MI 1.5 mm^2 2 core twin twisted cable	482	m		
Accessories for electrical services				
Accessories				
Call points				
Breakglass unit	9	nr		
Sounders				
Electronic	10	nr		
Flashing beacon type	6	nr		
Interface units, complete	2	nr		
	TO COLLECTION			

Figure 6.6   Bill of quantities

We would need to take into consideration works carried out at high level as these will take longer to do. We can fix a 3 m length of 100 mm × 100 mm trunking when standing at ground level quicker than we can using a working platform. The installation using a working platform is quicker than using a hydraulic lift, which in turn may be quicker than using a scaffold.

The standard rate for installing equipment does make allowance for working at different heights. As an example if we consider a high level run of trunking passing above structural beams we may allow for the height in our rate and would normally carry out the operation using a mobile tower. Because of the drop beams the operation will take longer as the top section of the tower will need to be dismantled each time a beam is encountered.

We can note this as one item for consideration and apply a factor to the standard rate to allow for this. It should be highlighted during the post estimate considerations prior to the figure for the tender being finalised.

There are other matters relating to the execution of the work that may have a bearing on the time taken or the cost of completing the contract. The best way to establish this is to consider:

- what the work is
- where the work is
- how long will it take to complete.

Let's look at some of the typical factors and their effect on the cost of the work.

Location

Is the site close to our own base? If not, but it is within commuting distance, we may incur the cost of additional travelling time. If it is beyond commuting distance we may be involved in the lodging expenses and allowances which are payable in such circumstances. Alternatively we may need to consider the use of local labour to carry out the work.

Is the site easily accessible by road, rail etc.? Deliveries will have to be made to site and special arrangements will involve additional cost. It is not always the remote location that will attract such problems. In the centre of many large cities there are restrictions upon the access of goods vehicles. Local parking and waiting restrictions can mean either special permits, or deliveries being made out of hours. All of these will attract additional costs which must be considered.

Will we need to have increased storage and security due to the location of the site? If the company uses its base as a store for materials then can it still be used in this instance? Large quantities of materials may need to be stored for short periods to accommodate the work's programme. We would stagger the delivery of material to minimise the quantity on site at any time. If, however, we are operating in several areas at the same time storage could create a problem.

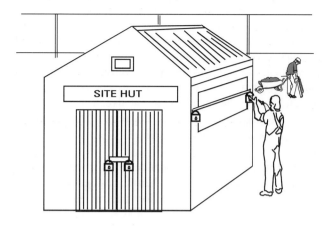

Figure 6.7   *Increased security may be needed due to location.*

Are there full services available from the public utility companies or do we need to provide these amenities? The cost of operating generators, portable toilets and providing water may be high over the period of the contract. An alternative is to have temporary services installed but the cost of these will depend upon the location and the availability from the relevant authorities.

Is the work site in a sensitive area and are we likely to encounter delays due to the normal operation of the site or its surroundings? New works carried out on existing sites and refurbishment contracts can be susceptible to disruption. This is generally caused by the daily operations of the existing occupants.

A further consideration must be for disruption from other sources. If we have a contract in a "politically sensitive" location, such as a main line railway station, we may suffer disruption from security alerts and the like. We are unlikely to be able to make an assessment of the disruption from such instances. However it is worth investigating what, if any, provision has been made within the contract to cover such eventualities.

Duration

What is the contract programme? If we are to complete on time will it be necessary for work to be done out of normal working hours? If we are to pay overtime the productivity to payment ratio can be as low as 50% of that for normal working hours and is something to which we should give serious consideration.

Overtime working may not be due to contract duration. It may be that there are certain activities that can only be carried out at night or over the weekend. We should be able to identify these requirements from the contract documents. However we must use our experience to identify any such areas at an early stage.

Will the contract run through the winter months and are there any activities programmed for this period that are weather dependent? In most well organised contracts any weather dependent activities are usually programmed during the summer months whenever possible. Whilst this does not guarantee the conditions, it does give the best chance of success.

For example, suppose we are programmed to install cables in an unheated building during the winter. It may be that the temperature will fall too low to allow that activity to be carried out. Similarly excavation works and cable laying to outside lighting, lamp standards etc. will be affected by the weather conditions.

Remember
PVC cables should not be installed at temperatures below 5 °C as the insulation is too brittle and is liable to crack.

At the other extreme in a building with a metal roof we may find temperatures during the summer too high. The installation of plastic conduits for example should not be carried out if the temperature is above 60 °F (AA6).

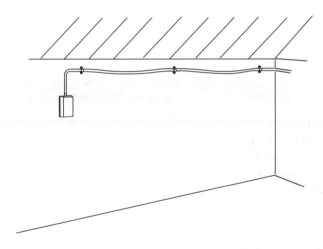

Figure 6.8

Once the building construction is complete the ambient temperature may be controlled to within the limits required for these materials. Until this is done temperature extremes both above and below normal could affect our installation work. Most contracts do not allow contractors to claim for additional cost due to weather conditions, these being classed as Acts of God.

Site measurements

If we are tendering for work in an existing building, we may be invited to attend the site to take measurements and view the working environment. We must make sure that accurate details are taken and recorded during any such visits. These will be required for the tender process and for reference if our tender bid is successful

Particular requirements

Some larger tenders make reference to sums that are to be included within the tender sum, these are referred to as
• Prime Cost (PC)
and
• Provisional Sum (PS)
and these are often included to cover work in a forthcoming project for which the full details are unconfirmed when the tender is sent out. In some cases this is due to an architectural detail (a specially built luminaire, the detail and cost of which has not been finalised). It may however be as the result of a technical requirement that cannot be resolved until work has commenced (the suitability of the existing Fire Alarm MI Cable installation for a new Fire Alarm system).

There is a distinct difference between a Prime Cost and a Provisional Sum which must be clearly understood, so we shall consider each in turn.

Prime cost

This is a sum of money which may be expended later on the purchase or supply of a particular item of equipment, say a new standby generator. The contractor will only be paid the trade price of the goods supplied under a Prime Cost sum and the client is entitled to ask for proof of the contractor's expenditure.

If we wish to recover handling charges and/or make a profit on goods supplied we must include this within the tender. Some tender build up sheets allow space for this to be declared. Should this not be the case, or we wish to recover more profit than it would be advisable to show, we must allow for this elsewhere. This is usually done by inflating one or more of the other sums making up the tender. We could, for example, add a small increase to items that are required in large numbers in order to make up a larger additional sum for our Prime Cost item.

Provisional sum

This is a sum that will cover the installation of a particular part of the work. Our previous example of the fire alarm cables could be covered this way. If the existing cable is not suitable for the new system then the supply installation of the cable could be covered by a Provisional Sum.

It is not uncommon for the contractor to be asked to quote for a Provisional Sum. As the Provisional Sum is a "gross" figure then the contractor will include all costs and profit in the quotation for the Provisional Sum. It would not be necessary in that case to include overheads and profit elsewhere in the tender.

Having considered all these factors, we can put together our cost for carrying out the work. The quotations from suppliers should have arrived and may be added to our total cost for materials and labour to give an overall cost.

Other considerations

We must then add the overhead and establishment costs.

These will include the costs for the company offices, vehicles and support staff. These costs are often added as a percentage to the total cost figure and represent the overheads.

The establishment costs will include those of setting up on site and providing the tools and plant the workforce will need to carry out the work. It is not uncommon for these to be spread over the various items that make up the tender.

Management considerations

Having arrived at our total estimated cost we must now apply the management functions to arrive at the final tender figure. These will typically include the increase or decrease in the estimated figure based on

- the specific contract conditions
- any special requirements with a high risk element, this could be financial, environmental or a particular activity
- any particular market forces or client relationship that may require an adjustment to the overall cost. A typical example is the possibility of further work for the same client. Should any new client be a likely source of good continued work a downward adjustment may be considered prudent in order to secure the initial work.

This is not a preclusive list, it only serves to give some indication of the type of considerations given.

We then have to add the profit to the final cost. This is most often done as a percentage of the estimated figure added to the final cost. The exact percentage is the final management function before the tender can be completed.

The final step is to prepare the tender documents and return them to the client. Included with the tender bid should be;

- details of the proposed programme for the work
- any particular exclusions or assumptions made
- details of any sub-contractors proposed
- savings which could be offered by the use of alternative manufacturers, suppliers, changed working practice and the like.

We must also notify the client, at this stage, if there are any deviations from the contract conditions.

Remember
The art of good estimation is to arrive at the most accurate estimate for the true cost of carrying out the work required. Whilst additional monies can be included to cover any unexpected costs this is likely to produce a figure too high to win the job. A figure too low may result in the company making a loss on the works should the tender be accepted.

A good estimator will try to clarify any uncertainties, detail any assumptions or exclusions and try to obtain the best possible discounts from suppliers and manufacturers.

It is the estimator's skill that will give the company the most competitive advantage, thus securing work, growth and profit for the company.

PROJECT

Now we have completed Chapter 6 of the module we should consider part 5b of the project. In this we shall need to produce a distribution diagram for the installation. This will require us to carry out a "take-off" and costing exercise so we must make sure we have access to the necessary cost details.

7

Planning

In this chapter we shall look at the requirements for the planning, programming and scheduling of work. For a small domestic installation we need to consider obtaining material, arranging access, and planning when we will start our next job.

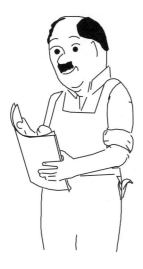

Figure 7.1

On completion of this chapter you should be able to:

- state reasons for planning and programming works
- prepare and use bar charts
- interpret networks and critical path
- assess use of resources to achieve programme needs
- demonstrate need to co-ordinate requirements with other trades and client
- detail needs for phased deliveries throughout the contract to suit site conditions and progress
- explain the need for planned utilization of resources

Planning a programme

The materials for a domestic rewire are generally available from stock at the wholesalers and any "specials" are usually a fairly short delivery period. We have established during our tender process the number of operatives needed to carry out the work and the time scale we anticipate for the job. The actual access and start dates are usually arranged to suit our client's requirements and our availability.

In planning our small domestic installation we have considered all the activities necessary to produce a basic programme for our rewire. The chart in Figure 7.2 illustrates a simple logic sequence for carrying out any installation work.

ORDER RECEIVED

⇩

ORDER MATERIAL

⇩

ORGANISE LABOUR

⇩

DELIVER 1ST FIX MATERIAL AND LABOUR

⇩

1ST FIX INSTALLATION

⇩

DELIVER 2ND FIX MATERIAL

⇩

2ND FIX INSTALLATION

⇩

INSPECT AND TEST INSTALLATION

⇩

COMPLETE THE DOCUMENTATION

⇩

HAND OVER JOB

Figure 7.2 Simple logic sequence

We shall briefly consider the items in Figure 7.2 and the sequence of events to help us appreciate the logic and programming for more complex works. For the purpose of this exercise we shall assume that the work will be carried out in a vacant property and will have definite first and second fix stages.

As soon as we are notified that we have secured the work we must refer to our quotation (tender) and arrange our activities along the following guidelines:

- Are there any items which are unusual? These are normally of a specific type or manufacturer. In this case they are likely to be particular fittings or accessories requested by the client. Any such items may have a longer than normal delivery period and so should be placed on order straight away and the earliest delivery date obtained. This will ensure the delivery whilst the job is in progress.
- We shall also have to check the level of manpower required to carry out the work and we need to know when they will be available to start this work. From this and our delivery dates we can determine our earliest start date on site.
- At this stage we need to consider the time scale for the work. We shall need to arrange our next job for the labour team to start when this one finishes.
- We can now arrange, with our client, the actual start date.
- For a domestic rewire we can usually collect the required materials from the wholesaler in one visit. This material can then be stored on site and used as required. We can ensure that labour and first fix materials arrive on site to begin work on the appointed date.
- Carry out first fix
- The second fix materials should arrive on site just in advance of the second fix beginning. Failure to do this will result in loss of productivity and add additional cost to the job.
- Carry out second fix work
- Complete Inspection and Testing of the installation
- Produce the Certification documents and hand over the completed Documentation and the work to the client.

This is an over-simplified procedure but the intention is to provide us with an insight into the logic approach that will be required when we consider more complex projects.

Before we move on we can use the list to produce a bar chart to demonstrate our progress. To keep this simple each activity must be complete before the next activity begins and we will then finish up with a chart similar to that shown in Figure 7.3.

Remember
It costs a lot more in time and money to return to a job when it's complete to carry out a simple task like installing a light fitting than it does while we are on site carrying out work.

We are often not given the luxury of planning our work to our own time-scale. It is far more common for us to be given the start and finish dates for a project. Within this we will be allocated time periods to carry out our activities. In such cases we need to plan our material deliveries and labour levels very carefully, in order to complete the work economically and within the time-scale allowed.

OPERATION DESCRIPTION	DAY NO.								
	1	2	3	4	5	6	7	8	9
1ST FIX INSTALLATION	████	████	████						
2ND FIX INSTALLATION				████	████	████			
INSPECTION AND TEST							████		
HANDOVER								H/O ◇	
DELIVER 1ST FIX	◇								
DELIVER 2ND FIX			◇						

Figure 7.3   Simple bar chart

Now suppose we are engaged to carry out the electrical installation on a new housing estate. There will be a number of factors that need to be considered which are beyond our control.

We are dependent upon the building progress to determine when we can carry out our activities. Likewise the builder will depend upon our progress to allow the next operation to begin. The builder will also need to co-ordinate the activities of other trades involved in the construction and finishing stages of the project. This programming and scheduling requirement will apply to all contracts irrespective of size.

As we become involved in contracts for larger projects we will enter into a formal arrangement (contract) with our client. When we are sub-contracting to another company, a builder for example, it is the builder who is our client. When we are employed to carry out any work our client is the person, or company, that engages our services. They in turn may be engaged by their own client.

In many cases the contract conditions include the payment of Liquidated Damages for delayed completion. These are usually calculated on the losses that the principal client is likely to suffer should the contract be delayed. These may bear no relationship to the actual value of the contract. In the case of a large project, such as a shopping precinct for example, the damages may be set at many thousands of pounds per day.

The Builder, or Main Contractor, will usually include these conditions within the sub contract. We could then be liable to pay the Liquidated Damages for delay should we be responsible for the failure to complete on time. We must therefore be sure that we can carry out the works in accordance with any programme produced.

Typical contract relationships are shown in Figure 7.4.

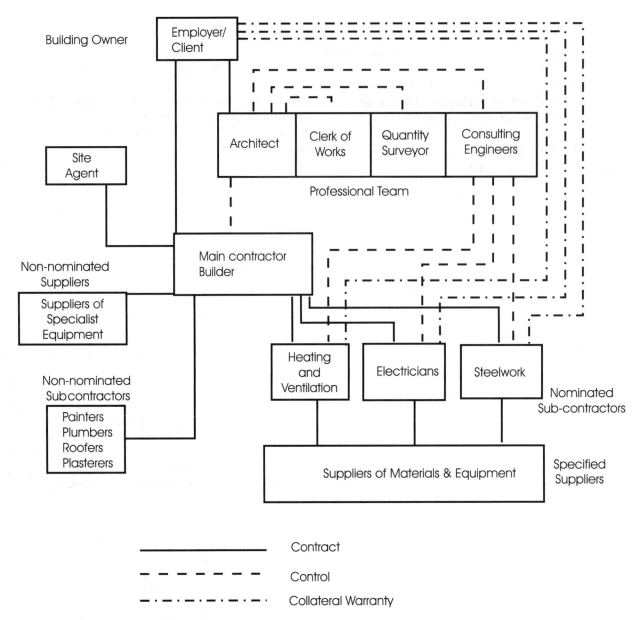

————————		Contract
— — — — — —		Control
— · — · — · — ·		Collateral Warranty

Figure 7.4   Typical contract relationships

Contract relationships

We have used a few terms which you may not be familiar with so perhaps a brief résumé of who's who would be useful.

- **Main Contractor:** This is the contractor who is awarded the contract for the entire project. In many cases it is the building contractor who carries out the construction of the project.

- **Subcontractor:** This is a term applied to companies who take on specific parts of a larger contract. These are usually for the Main Contractor and often involve the "Specialist Trades" such as electrical, plumbing and ceiling erection. These companies enter a sub-contract with the main contractor which will have its own terms and conditions. These terms and conditions will usually incorporate those of the main contract. The terms of the subcontract sometimes allow for the further subcontracting of work, subject to agreement. We may, for example, wish to engage the services of a specialist Public Address or Fire Alarm contractor, who would then be subcontracted to us.

Remember

The delay is not necessarily caused by the activities which remain outstanding at the end of the programme time. For example, imagine we fail to complete our first fix activities in one area during the early stages of the job. As a result the following trades could not start to their programme. Although all the activities were completed in their allotted times, the time lost was never recovered. The Main Contractor would then show that the reason for the works failing to meet the completion date was entirely due to our failure to keep to programme in the early stages of the contract.

- **Management Contractor:** This is a company engaged by the client, which carries out no construction at all. Their sole function, as the name suggests, is to manage the project. They are responsible for letting contracts for each activity on the project, programming and co-ordinating the works of all trades and controlling the payment of the contractors. They are paid, as the job progresses, directly by the client.
- **Builder:** Within the confines of this text the builder is a term used to describe the function of a construction company that is also the Main Contractor for our projects. In some cases the Builder may be the developer and is therefore the principal client as well as the main contractor.

At the time of tender a proposed programme is usually issued to indicate the time-scale of the project. It will normally indicate the "time slots" provisionally allocated to each trade for their activities.

It is quite common for the main or management contractor to produce the overall programme for the contract. In order to do this they will ask each individual contractor to produce a programme for their particular activities. These programmes are then used to produce the overall programme for the project. It is not unreasonable to expect that our working programme should be close to the time slots indicated on the tender programme.

Let's take a look at the considerations for programming our works. We know when we are to start work and we must ensure that tools, plant, materials and facilities are available on that day. If we consult our tender details we can determine how long each activity should take to complete and we can then plan the periods in a logical sequence in the programme.

This, we know, can be shown as a simple line with the duration of each activity shown running sequentially as in Figure 7.5. Now it is very unlikely that the work will be carried out in this manner as we may begin second fixing in one area of the installation before first fix is complete. Equally we may begin first fixing lighting and power together and so on.

To cater for these variations we can produce a simple bar chart to show the sequencing of events. The chart shown in Figure 7.6 gives a more practical interpretation to the activity sequence. We can see that in several cases work is being carried out at more than one work face at the same time. On larger projects this chart will become considerably more involved and many more activities will overlap.

This would be a typical proposed working programme which we would forward to the main contractor. Whilst this contains all the detail relevant to our activities there are the other construction trades to be considered. For example the plumbing and air conditioning first fix also needs to take place before the plastering and ceiling erection. The builder, by nature of the construction process, may be restricted to progressing the works in a specific order and at a predetermined rate due to curing times etc. All these factors will have an effect on the final working programme.

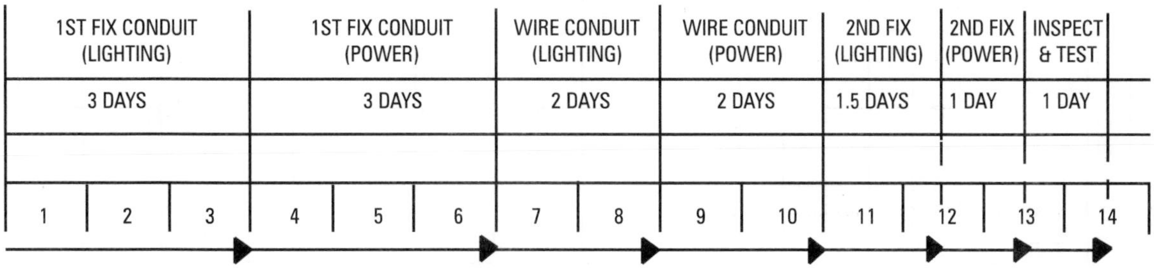

1ST FIX CONDUIT (LIGHTING)	1ST FIX CONDUIT (POWER)	WIRE CONDUIT (LIGHTING)	WIRE CONDUIT (POWER)	2ND FIX (LIGHTING)	2ND FIX (POWER)	INSPECT & TEST
3 DAYS	3 DAYS	2 DAYS	2 DAYS	1.5 DAYS	1 DAY	1 DAY

Figure 7.5   Linear programme

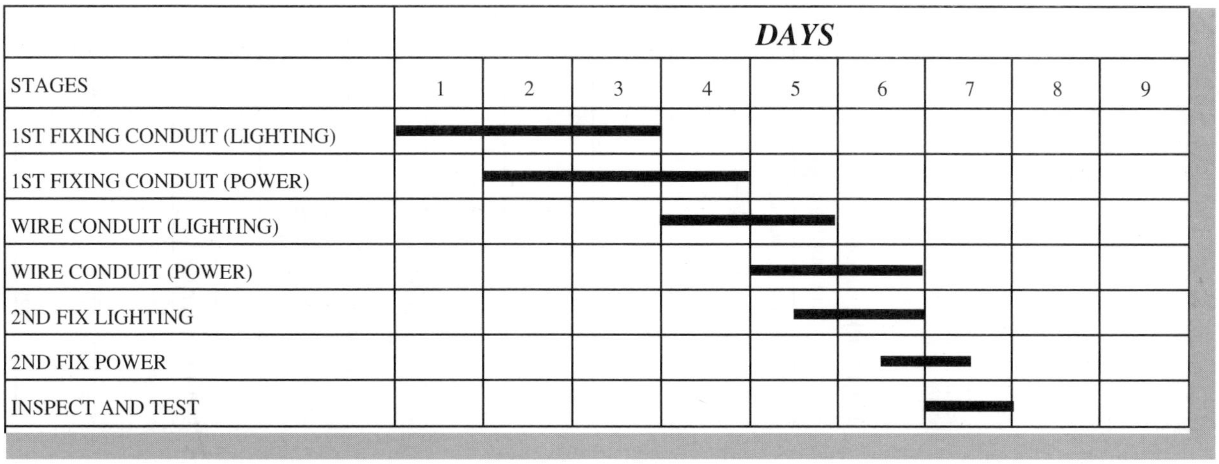

	DAYS								
STAGES	1	2	3	4	5	6	7	8	9
1ST FIXING CONDUIT (LIGHTING)									
1ST FIXING CONDUIT (POWER)									
WIRE CONDUIT (LIGHTING)									
WIRE CONDUIT (POWER)									
2ND FIX LIGHTING									
2ND FIX POWER									
INSPECT AND TEST									

Figure 7.6   Bar chart

Eventually we will have our construction programme with our working periods detailed, an extract from a typical programme is shown in Figure 7.7. We need to ensure that the necessary materials are available for each stage and arrange for delivery to site accordingly. Any long delivery items should have been identified at tender stage and placed on order as soon as the contract was awarded.

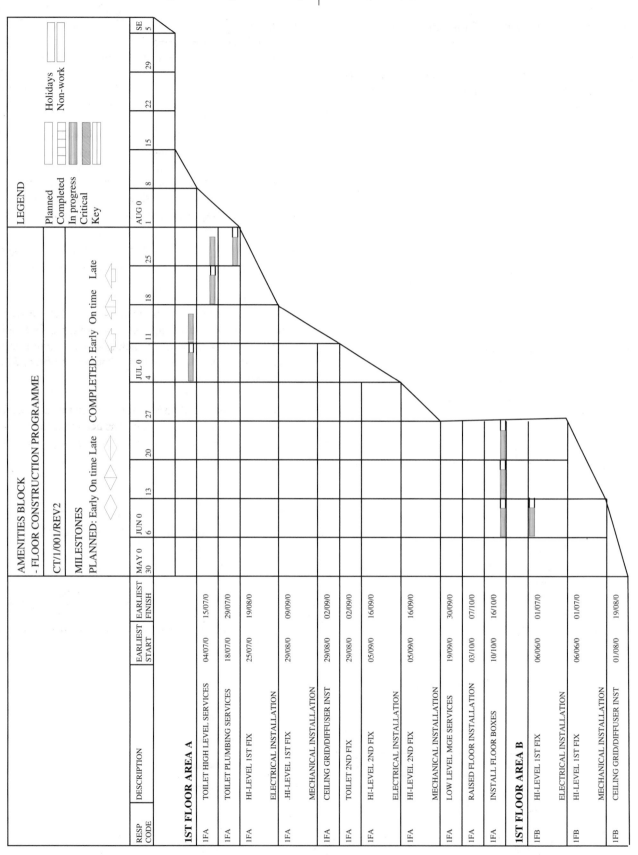

Figure 7.7   Typical bar chart programme

Critical path network

As construction projects become increasingly more complex, the planning and the execution for completion at a given time become more important. One way in which this is achieved is by the production of Critical Path Networks.

The principal functions of these networks are to:

- ensure that all information, plant, labour and materials are available when they are required during the project
- avoid excessive and costly peak demands for labour and plant
- make sure that the project is completed on time

With every contract there will be a "Critical Path" and the activities which make up the Critical Path must be carefully monitored to ensure completion on time. We shall consider a Critical Path for a fairly simple operation as shown in Figures 7.8, 7.9 and 7.10.

In the critical path in Figure 7.8, the activities considered are linear and the completion of each activity is vital to completion of the job. The effect of changing this linear criteria is shown in Figure 7.9.

A typical network is, at first sight, somewhat complex but it is really no more involved than a regular bar chart, it's just not as user friendly. In a typical critical path network we show each task as an arrow with the base of the arrow as the start and the point as the finish.

A critical path network is able to show diverging and converging activities in a way not possible with a bar chart. All the activities which must be carried out in sequence and the restraints placed on one task by another are considered.

The network provides a means of calculating the time to complete the project, based on the time taken for each task and their sequential relationships.

The route through the network which takes the longest time to complete is the critical path. Should the tasks or activities along this path be disrupted or take longer to complete then the project will be delayed and will not finish on time.

To produce a network we first need a list of all the tasks we must carry out to complete our work, as we did for our simple programme. When we do this it is a good idea to list the time period, the plant and equipment, the labour level needed as well as any special materials for each of our tasks.

To draw the network we must now place the arrows in position, head to tail considering

- the tasks which must be completed before another task can begin
- the tasks which cannot begin until another task is complete
- the tasks which can be carried out at the same time as other tasks.

Once we have produced our network and all the arrows are placed it is usual to place numbers for identification to all the events at the tail and head of each arrow. By convention the numbers at the tail of the arrow are lower than the numbers at the head.

Each activity starts and finishes at some point in time and these are referred to as events. An event has no time duration, as it is simply the point in time when an operation begins or ends.

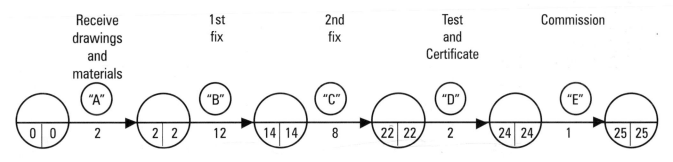

Figure 7.8   Simple network

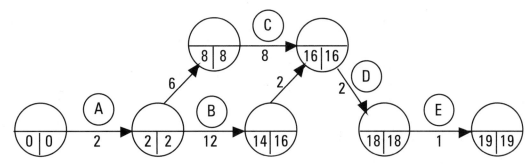

Figure 7.9   This is the network we get if the 2nd fix can begin 6 days after the start of the 1st fix, and the 1st fix must finish 2 days before the end of the 2nd fix.

155

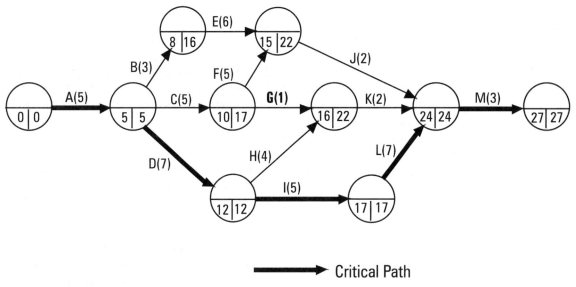

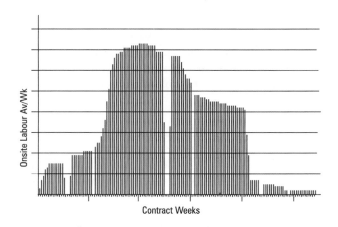

Critical Path

Figure 7.10   Typical network with critical path highlighted

However, each event does represent a particular point of time within the programme, often referred to as milestones, which mark the end of one operation and the beginning of another. Each event is usually given a unique number and these are used as the identification of progress against the critical path.

By using the times from our estimate for each activity we can determine the time likely to be required to complete each task. The time unit used is normally based on a 5 day working week with each day equivalent to 0.2 of a week.

The network provides a useful management tool with which to monitor and control the project as it progresses, it is not however, in the most easily read format. This is usually overcome by the production of a bar chart programme which accurately reflects the network as a series of bars related to each activity and duration.

A typical bar chart programme is shown in Figure 7.6. It is quite common for the event numbers from the critical path to be shown as milestones on this programme.

Resources management

In addition to establishing the programme and critical path, the network allows us to ensure the most cost effective and economic use of materials, labour and plant throughout the project.

We will often be asked for a labour histogram which shows the labour required at each stage of the job. The labour histogram may be represented as a line graph, or more commonly as a vertical bar chart.

It is usual to produce a histogram for each trade involved in the construction process. These may then be summed to determine the total labour present on site at any time. A typical labour histogram is shown in Figure 7.11 with a summary resource chart in Figure 7.12.

156

Figure 7.11   Labour histogram

It is important, during the planning stages of any job, that we give serious consideration to the allocation of labour. It is a good working practice to keep a group of employees involved with the same project throughout.

When any employee arrives at a new job there is a period of familiarisation. They need to know where:
- everything is kept
- where each work face is located
- what is happening in each area
- who is in charge and so on.

During this period the employee is not fully productive and the productivity of fellow workers falls as result of their advising a new member of the team.

Project Name: Resource Chart	Programme No.
CONTRACT WEEK NO.	PLANNED RESOURCE
1	3
2	7
3	9
4	12
5	13
6	15
7	15
8	15
9	15
10	15
11	15
12	15
13	15
14	8
15	0
16	0
17	9
18	19
19	19
20	19
21	19
22	19
23	19
24	21
25	
26	
27	

Figure 7.12   Resource chart

The best balance for productivity and moral is to maintain a fairly constant level of labour throughout, unless this is totally uneconomical because of the rate at which areas become available. We must then use our skill and judgement to make the best use of the labour available and keep increases and decreases in the level to a minimum.

As the level of labour rises and falls so too will the amount of plant and equipment required. The effect of this will be a rise in the cost of plant (we cannot negotiate good terms for spasmodic hire of plant). The cost of purchasing equipment which will then remain idle for periods of time is not economic. It may also have no redeemable value at the end of the contract and must be funded from the job overheads.

A good example of effective resource management would be the lifting of equipment to a roof plant room by crane. We may have a number of large control panels and a distribution board to be installed on the roof. The mechanical contractor will have air handling equipment, large pieces of pre-formed ductwork and the like to go onto the roof. It is likely that the builder will have construction material to go to roof level as well.

In order to carry out this operation we may have to request the closure of roads around the site and hire a mobile crane for lifting. By careful planning of deliveries and availability of labour the lifting exercise for all trades could be carried out during one period. This will save costs for road closures, crane hire and labour to carry out the work.

On a smaller scale imagine we have a number of sub mains cables to terminate which require the use of a particular crimping tool. If we can arrange to have all the sub main cables glanded off awaiting final connection we may be able to hire the crimping tool for a single day. This is much more economic than having it on site for several weeks and only used for an hour a week.

There are many ways in which we can, by the careful planning and utilisation of resources, make our contract more cost effective. It is a common mistake to underestimate the cost of small plant and equipment over the duration of a contract.

If we double the number of men on the site for a short period of time we shall need twice as many sets of steps, 110 V power tools, etc. There are also the welfare facilities to be taken into account such as mess rooms, toilets and drying rooms. These will all need to be provided even if the increase in manpower is only for a short period.

Try this
Consider the following activities on a large site and suggest ways in which each could be made cost effective.
- The unloading, storing and transporting around the site of materials
- Keeping the work face clear and tidy
- The Inspection, Testing and Certification of work carried out (this will include documentation such as the record drawings)

These are some of the steps that could be taken to make these activities more cost effective:

Materials

- Arrange for vehicular access up to the storage area
- Whenever possible arrange for deliveries at particular times, for example between 10.00 and 12.00 each day
- If material can be delivered direct to location on site arrange for this to take place as it saves relocation time.
- Make sure lifting tackle, rollers, supports, cable jacks etc. are available when a delivery arrives.
- Arrange deliveries to coincide with the requirements for installation to programme.
- Use labourers to unload, locate and position materials and equipment. Their cost to the contract is lower than those of skilled labour.
- Arrange for suppliers of major items to off load and position their equipment as part of the order. Preferably at no additional cost.

Keeping the work face clear and tidy

- Use unskilled labour for the removal of rubbish
- Arrange a suitable location of materials and tools for use at the work site. The use of reinforced, lockable site boxes in which to keep equipment allows us to reduce the time moving gear from the store to the work face each day.
- Allow eating and drinking in designated areas only.
- Arrange for regular clean up times, for example if the site closes at 18.00 arrange for labourers to start cleaning at 17.00 leaving the site clear overnight and ready for work to start the next day.

The Inspection, Testing, Certification and Documentation

- Carry out inspection and testing for each area or distribution board as they become available.
- Consider the use of a special team to carry out all the Inspection, Testing and Certification. This will help to standardise the approach and the team will be able to familiarise themselves with the whole job.
- Carry out any remedial work highlighted by the testing team before the area is vacated.
- Complete and hand over, where possible, individual areas once they are complete. This will allow the main contractor to take responsibility for the areas and minimise our risk of claims for loss or damage.
- The testing team will hold and be responsible for their instruments which will only be used for final testing. These will need to be kept in calibration and can be used to verify results if required.
- Use standard forms and test results sheets throughout and have the testing teams verify the circuit charts for the distribution boards so that these may be produced and fitted in their final form prior to completion.

- Ensure that area supervisors maintain marked up as fitted drawings. These should be revised on a daily basis and can then be passed to the draughtsmen for the production of the fair copy record drawing as soon as the area is complete.
- The supervisor should also complete the inspection during construction records on a regular basis.
- Copies of the forms used for certification, particularly the schedules of items inspected and tested, should be used during the working phases of the installation. The details on these can then be readily transferred to the final Electrical Installation Certificate for issue to the client.

PROJECT

Now we have completed Chapter 7 of the module we should consider parts 6a, and 6b of the project. For this we shall need to produce a simple bar chart and labour histogram for one of the activities so make sure you have the necessary drawing equipment and materials available.

8

Administration

In this chapter we shall look at the site administration and record keeping for our projects.

The construction industry has become ever more aware of the cost and legal implications of carrying out work. As a result the documentation and paperwork involved in the running of a job increases. Many companies are now seeking Quality Assurance Accreditation to BS EN: ISO 9000. This requires that the company procedures for carrying out its business are detailed and standardised throughout the company.

As we are involved in a variety of paperwork, in this chapter we shall consider this aspect in more detail. Some of the items will be familiar to you and others you may not have encountered, but they all have their own particular function.

Figure 8.1

On completion of this chapter you should be able to:

◆ state records that need to be maintained on site during progress of work
◆ list documentation issued and received by an electrical contractor during progress of works
◆ complete necessary documentation for employer and client to discern, prove and advise of departures from the installation as tendered
◆ demonstrate an understanding of British Standard electrical symbols used for drawings
◆ state the reasons why record drawings are required

Time sheets

Most companies use a time sheet of some kind, many producing a sheet with their own particular requirements. The time sheet provides the company with a written statement on the activities of the employee. A typical time sheet is shown in Figure 8.2.

DOUGHTON BROS.
Electrical Contractors
TIME SHEET

Name _____

Week ending _____

	Job No.	Time Started	Time Finished	Total for day	Travelling Time	Mileage and fares
Sun						
Mon						
Tues						
Wed						
Thurs						
Fri						
Sat						
			Total			

Operative's signature _____ Date _____

Foreman's signature _____ Date _____

Figure 8.2   Time sheet

Whilst the detail shown on this example is not exhaustive, it does represent the minimum required by most companies. These are normally completed on a daily basis and handed to the electrical foreman at the end of each week for countersigning. They are then sent to the office for processing. As far as the electricians are concerned the time sheet is the document that determines the pay they will receive.

As far as the company is concerned it is the document that is used to calculate and substantiate the wages paid. This information is then used to determine the tax deducted, national insurance, any pension contributions and the like. Should a dispute arise over the time claimed for additional work etc, a time sheet, duly completed and countersigned, may be accepted as evidence. This could then help to substantiate the numbers of personnel on site at any time.

Day work sheets

Day Work sheets are used by many companies to record day to day activities of employees on site. In the majority of cases they are used to record additional or variation work that is required.

Let us consider an example where we have installed a run of trunking in the correct position and in accordance with the programme. The main contractor has a piece of equipment that was located on another part of the site and is now to be removed. Our trunking crosses the only route out of the building, and whilst erected at the correct height, it is too low to allow the equipment through.

The Main Contractor approaches us to remove a section of trunking to enable the equipment to be removed from site. As the work needs to be carried out fairly quickly, and is an activity that would not have been allowed for and therefore not priced in our tender bid, we are asked to carry this out on a Day Work basis.

Doughton Bros. Electrical Contractors

DAY WORK SHEET

Customer: Address:

Job No:

Description of work carried out:

Date	No. of men	Names of operatives	Time started	Time finished	Total for day	Travelling time	Mileage and fares	Notes
		Totals						

MATERIALS

Quantity	Catalogue no.	Description	For office use

Foreman's signature   Date

Customer's signature

Figure 8.3   Typical day work sheet

This means that the Main Contractor's supervisor will record the time and number of our operatives working to remove and replace the trunking. The operatives fill in a Day Work sheet each day, this is checked and signed by the electrical foreman. This must usually be presented to the Main Contractor for counter signing usually within 24 hours.

Providing the Main Contractor agrees with the time taken and the number of operatives involved the Day Work sheet will be countersigned. Payment will eventually be made at the rate agreed within the contract.

A typical Day Work sheet is shown in Figure 8.3 and as we can see there is a provision for recording any materials used whilst carrying out the work. Many companies also record plant and equipment required. We may, for example, have to carry high pressure testing of a High Voltage Cable suspected of damage. This is not likely to be the sort of equipment carried by the average contractor and would therefore need to be hired.

We may also use a Day Work sheet to record details of small jobbing works carried out, such as minor domestic repairs.

Schedule of rate

It is common for us to be asked to produce a schedule of rate for all the activities covered by our tender. This is usually required soon after the contract is awarded. The Main Contractor will then pay for any additional activities against this rate.

So if we are required to install an additional 50 metres of 25 mm black enamel conduit and this takes two men all day we duly record this on the Day Work sheet and it is countersigned by the Main Contractor. We shall only be paid the rate in our tender for the installation of 25mm conduit irrespective of which cost is greater unless unusual circumstances can be proven.

Remember
A Day Work Sheet can be used as a record of activities carried out on a large contract and the details contained must be correct.

Payment to the company will only be made at an agreed hourly rate if the work involves activities **NOT** included in our tender.

Site instructions

Whilst we are working on site, it is inevitable that we will receive instructions that will vary our work or method of operation. Typically we may be asked to:
- Install additional materials, plant or equipment
- Omit material, plant or equipment
- Change the material, plant or equipment to be installed
- Vary our working hours, times on site, weekend working etc.
- Change our programme of work
- Alter the position or location of material, plant or equipment

It is important that all the paperwork associated with instructions is recorded and stored in a secure location. We will generally have a requirement within our contract to respond to an instruction within a prescribed period, usually 7 days. The response should inform the client of any cost or programme implications resulting from the instruction.

Any instruction which changes what we are to do, what we are to do it with, or how we do it must be issued by someone with the authority to issue such instructions.

Most site instructions will be issued in connection with items of work for which the Main Contractor is responsible. Changes to the work we are to carry out are usually instructed by way of Variation Order or Architect's Instruction, these will be dealt with later in this chapter.

The general rule is that we can only be instructed to make changes by the person or company with whom we are contracted to do the work. If we are contracted to a Main Contractor the client, or the client's representative or consultant will instruct the Main Contractor, who will in turn instruct us. This can, of course, be a different arrangement according to the precise wording of the contract.

This does not mean that we cannot discuss the required changes with the other parties. However, until the official instruction is issued there will be no contractual agreement in place for any variation. This means that there is no requirement for payment to us and that we could be asked to revert back to the original requirements for no additional payment.

It is for this reason that many companies operate a policy that no change will be initiated until an official instruction is received.

Variation Orders and Architect's Instructions

On most contracts there will be a standard variation sheet issued for all official Variation Orders (VO). As these are issued each will be given an individual number. All parties involved will refer to the works covered against this individual number.

A Variation Order is usually issued when a Management Contractor or consultant is responsible for controlling the changes and financial expenditure of a project. Without this instruction no variation work is authorised and will not be paid for.

The official issue of an instruction is often done by Architect's Instruction (A.I.). The client will often engage an architect to be responsible for a total project. This will include the control of all expenditure and the authorisation of changes to the original scheme.

The AI and the VO are almost the same, which one is issued usually depends on how the client has set up the contract. They should both be treated in the same manner.

The AI or VO may not carry any cost or programme implication but without it we will be obliged to install to the original requirements. If the AI or VO is not issued then we will not be paid for any uninstructed variations.

Upon receipt of an instruction we must ensure that we record the instruction number, the date it was received and a brief description of the content. We must then investigate what is required and advise the client, within the stipulated time period, of any financial or programme implications.

When we respond to an instruction in this way we must make it clear that this response is only an estimate of the final cost. It will depend on the nature and extent of the variation as to how accurate our estimate is going to be.

Confirmation of receipt of instruction

When any instruction is received, we must notify the Client's representative, this may be the Main or Management Contractor, it could be the consultant, or the client direct, of its receipt.

This is usually carried out automatically for official written instruction by the need to respond within the stated time period. There are some occasions however when this does not or cannot happen and there are instructions given other than through the official channels.

We are likely to receive instructions from the Main Contractor's representatives on site particularly where an operation has to be carried out quickly. These should take the form of a site issued, hand written instruction, but in many cases we will be asked verbally to carry out some work. We must notify the receipt of verbal instructions and exactly what it is we have been asked to do.

Many companies have a standard form to do this job and a typical example is shown in Figure 8.4.

Doughton Bros. Electrical Contractors

To: Contract No.
                                 Client Ref:
                                 Date:

Dear Sirs,
Re: CONFIRMATION OF RECEIPT OF INSTRUCTION
                   CONTRACT:
We acknowledge receipt from your representative ....
of instructions to vary our works as described below:

We confirm that in accordance with the terms and conditions of the contract between *(Co. name* ....) and *(Client name* ....) the above identified/described instruction(s) which have been received by *(Co. name* ....) will be valued in accordance with the terms and conditions of the contract. The above instruction may result in a variation to our contract price and programme.

Signature of Client's rep.
Issued by:

Figure 8.4   Confirmation of receipt of instruction

These are sent irrespective of how the instruction is received or from whom. If notification of the receipt is not refuted, usually within 14 days, then it is deemed that the instruction has been accepted and we are entitled to be paid for the work involved.

Drawings

Should the client wish to change part of the layout, either of the building or the services, or the design of part of the project is revised, then we will usually be issued with a revised drawing for the area concerned.

These drawing revisions are usually issued under an Architects Instruction and will have an AI number covering this issue. We may not be able to notify the cost and programme implications of these drawings due to the complexity of the changes or the number of drawings issued under the instruction.

We need to notify the receipt of the instruction and we need to keep a register of the drawing receipt. We must ensure that all site operatives and office personnel are using the latest drawings. A typical drawing issue register is shown in Figure 8.5 and a register of drawings received in Figure 8.6.

Doughton Bros. Electrical Contractors	Drawing Register Sheet No.																															

Contract No.

Job title

Section

Revision ⬭ Status

We have pleasure in enclosing for

1 Approval	**2** Construction	**3** Comments
4 Information	**5** Records	**6**

the following items

7 Drawings	**8** Sketches	**9** Annotated Prints

Drawing No.	Drawing Title	Issued for																			
		Item																			
		Received																			
File ref.	Distribution To	Issue Date																			

Drawings received

Signed: Date:

Figure 8.5   *Issue sheet*

Remember

We need to keep a register of all drawings and documents received from the client.

We must also keep a register of all drawings and documents which we issue. It is important that the latest revision is used for all on site work.

RECEIVED DRAWING REGISTER

FROM:
CONTRACT NO.
CONTRACT NAME:

Drawing No.	Drawing Title	Received date												

Figure 8.6   Received drawing register

Delivery records

Delivery records cover the delivery of materials and equipment to the site. There are a number of parts making up the complete set of records for ordering and receiving materials. For us to make sure that we receive the correct documentation and take the right action we must be aware of all the stages involved.

When any material is required, the company must first send a written order to the supplier. There are occasions when an order may be placed by telephone and the written part is referred to as a confirmation order. The order will include a specification of type, quantity, manufacturer and may also give delivery date, location etc. The company will retain a copy of the original order. In some instances you may have to raise an order for materials direct from site as a result of a variation that needs immediate action.

Materials delivered directly to the site will be accompanied by a delivery note. As the company representative on site this is the document that you are most likely to have dealings with.

The delivery note should state exactly what materials are being delivered to the site in this particular load. It may not contain the entire order originally placed by the company, this is often because not all the material is required at one time to minimise damage and loss during storage.

A typical delivery note is shown in Figure 8.7 and it should state:
- the name of the supplier
- to whom the goods are to be delivered
- the type, description and number of items to be delivered
- a statement as to the "state" of the order
- a space for the recipient to sign for the delivered goods
- a statement as to the time period allowed for claims for damaged goods

DELIVERY NOTE

DOUGHTON BROS.
Electrical Wholesalers

Delivery address _____
Invoice address _____

Order No. _____
Date _____

Qu.	Cat. No.	Description

Comments

Items received by: _____

Shortages and damage should be reported to us within 3 days of receipt.

Figure 8.7   Typical delivery note

The "state" of the order referred to may cover a number of conditions. As you will be signing for the receipt of goods it will be an advantage to be aware of these:

Incomplete order

This means that the delivery does not contain all the items that were on the original order. This may be due to the need to have materials delivered at different times, or that some of the order will come direct from the manufacturer. Another possibility is that the supplier is out of stock and that the order will be completed as soon as the supplier takes delivery of more stock. As far as we are concerned it informs us that there is another delivery of goods to come. This identifies that the shortfall in the delivery over what we expected is due to one of the above reasons and not to forgetting to order items in the first place.

Part of order

This is very similar to the above. The main difference being that in this case it is more probable that the order is incomplete due to a prearranged requirement rather than just being out of stock.

Completion order

This is the order that will finalise either of the previous two. It indicates to us that the original order placed has been completely filled on the receipt of these goods.

If the order is completed in one delivery then none of the above will be stated. Some suppliers will, however, stamp the order as complete.

So what is the procedure when goods are delivered on site? The first job is to ensure that they are unloaded and stored correctly. Whilst the goods are being unloaded, they should be checked off against the delivery note. At this stage we are only concerned with the quantities of goods and any obvious damage. Providing the goods delivered match those on the delivery note then you may sign for the items received.

If there are any items missing from the load that appear on the delivery note these should be recorded, on the delivery note, and signed by both yourself and the delivery driver. It may be an idea to notify the supplier by telephone in such cases to speed up the process of tracing and delivering the missing goods.

On the delivery note there will be a statement to the effect that goods damaged in transit must be notified to the supplier within a set time period, often 3 days. It is in our best interests to check the items delivered within that time and notify the supplier of any damage.

We must file the delivery note and keep it until such times as it is required for reference by us or the company.

Once the goods have been delivered, the supplier will send an invoice to the company. This is really a request for payment for the goods supplied. At this stage the company may wish to check the delivery notes against the invoice received to ensure that they have only been charged for goods they have received.

It is not uncommon, especially with suppliers of specialist equipment, to send the invoice along with the goods. In this case the invoice must be sent to the company office straight away. Many suppliers offer a discount for prompt payment and so having the invoice on site or even lost can result in loss of discount.

Remember
We should always check the items on the delivery note with the site copy of the original order. This is to establish that correct items have been supplied and note what remains outstanding.

Site diary

The site manager or site supervisor is responsible for keeping a site diary which should be completed during each day. It is used to record all the happenings on site, each working day. Once again the actual format of the site diary will vary from company to company but Figure 8.8 shows a typical arrangement.

Doughton Bros. Electrical Contractors		No.
SITE DIARY	Contract Date Contract No.	
PROGRESS – (Work done, location and if in advance, on schedule or behind)		
DELAYS – (Overdue materials, other contractors, etc.)		
MEMOS – To Depts. Managers, etc. (brief details of contents)		
MISCELLANEOUS – 1:V.O. Necessary; 2: Overtime (reasons); 3. Surveying (info.request); 4: Visitors 5: Other business		

PERSONNEL ON SITE: F/M C/H TECH AE ELECT APP JOINTERS LAB	WEATHER:	ACCIDENTS: Witness
Date	Signed	

DISTRIBUTION AND PRIORITY

Priority	Title	Name	Priority	Title	Name
	Divisional manager engineer			Contacts manager	

Figure 8.8   Site diary

The details usually required as a minimum are

- day, date and contract details, i.e. address, job number and client
- the name and designation of all our personnel on site (this may need to be recorded on a separate sheet, particularly where large numbers of operatives are involved, and a copy has to be given to the Main Contractor each day).
- telephone and Facsimile calls made and received
- deliveries received and goods returned

- site meetings held giving a reference as to the topic and who was in attendance (this is usually by company and reference to the minutes taken)
- instructions and drawings received and issued
- information requested
- any incidents, occurrences or accidents on site (with reference to any accident book entry)
- any disruption or completion/handover etc.

Whilst this list is not exhaustive, it does include most of the information that needs to be recorded. This document can prove to be most useful in the event of any dispute as to when things happened, who said what and other contentious items.

An experienced site manager will know instinctively what needs to be recorded because of the possible repercussions later. It is a most valuable and important record and should not be overlooked.

Reports

In addition to the usual reports on the progress and activities on site which are part of the effective site operation, it is also necessary for incidents occurring on site to be reported. The list of items required in the site diary does not leave a lot of space for recording events where there is much to record.

Many companies have a standard method of reports for these occasions and an example is shown in Figure 8.9. These reports are then individually numbered and reference can be made to the report number in the site diary and correspondence. In this way a brief comment can be recorded in the diary and and the detailed report can then be faxed, copied and distributed as necessary.

Doughton Bros. Electrical Contractors	Distribution		
		Action	Information
Job No.		Ref:	Date:
To: From: Subject:	Site: Reason: Authority		
REPORT ON OCCURRENCE/ACTION/REQUEST Numbered Items			Action by

Figure 8.9   Typical report sheet

Information and instruction required

During the course of the project we will find that there are items about which we do not have enough information. In such instances we need to request details from our client.

Most companies will have their own standard form or letter for this and we fill in the information that we require. It is essential that the request is clear and concise so that there is no doubt about what we wish to know.

There will also be occasions when we will need to seek an instruction to vary our work. One example would be where we need to divert our services around a particular building obstruction, say a supporting beam, that was not identified on the drawings.

Once again there will probably be a standard format for this and we must ensure that we state exactly what we require. Remember that the instruction must describe what we are required to do when it is issued. If we are able to do so, dependent on the urgency of the operation, we should notify the cost and programme implications of the request when it is submitted.

It is important when requesting information or instruction to state the latest date by which this needs to be received if we are to remain on programme. This must take account of ordering, delivery and installation constraints.

If a response is not received by that date, we shall need to notify the client of a delay to the programme. This will need to be monitored and the client advised on a regular basis if we are not to incur additional penalties for delay.

We must maintain a register of our requests and this should contain a minimum of

- reference number for the request
- date the request was made
- latest date for a response
- subject of the request
- date response received
- is the response satisfactory yes/no
- further action or request made (with the appropriate reference number)

This information can then be discussed at regular site meetings and the flow of information easily monitored.

Conditions of contract

Having already considered some of the documentation that we shall need to complete during the course of the work we should take a little time to review what is involved in accepting a contract.

If we agree to carry out a small job of work for a domestic consumer there is not usually any written contract. Due to the relatively small value of the work and the nature of the installation it is often not considered necessary. However, written contracts offer protection to both the customer and the contractor. Simple forms of domestic contract have been produced and proposed as part of the DETR's "Combating the Cowboy Builders-Working Group Report" published in 1999.

Figure 8.10   Draft contracts

167

Normally there is some discussion between the contractor and the client as to how the job will be carried out, the finish that is required and the equipment and accessories that will be used. Any dispute between the client and the contractor is either settled by discussion or taken through the legal system. Indeed the values of work involved in such disputes can often be sufficiently low to be dealt with in the small claims court.

With larger projects there is usually a fixed date for completion and the client, or the client's representative, will usually set down conditions related to:

- the execution of the works, and the associated responsibilities and liabilities
- the issue, receipt and actioning of instructions
- loss or damage
- responsibility for plant and equipment
- completion of the work and extension to time
- failure to complete on time
- payment
- disputes and arbitration
- fluctuation in labour, materials and tax

This list is not exhaustive and covers only the major items, many others are included in the contract documents but we shall not be considering them at this time.

The contract places constraints upon what we can and cannot do. The determination of the implications and effect of these terms and conditions is a highly specialised job. Any disagreement or dispute will need to be dealt with under contract law.

As the electrical work is usually subcontracted from another party, it is common for the work to be under one of a number of standard forms of subcontract used for these arrangements, typically Domestic Sub-Contract DOM/1 and the JCT 80. (Domestic in these terms does not mean "domestic" as applied to dwellings) Other forms of contract are available and used for works contracts, produced by parties such as Government Bodies and The Association of Consultant Architects.

Figure 8.11   Typical subcontract form DOM/1 reproduced by kind permission of CIP Ltd

We have already considered the requirements of the subcontract and the implications that this will have on our operations. When we accept the contract and its conditions, we will find that one of them refers to the documents issued for the purpose of providing the tender. It is important that we are aware of exactly what was issued for tender and any additional documents used in the tender process. As a general rule these will include a minimum of

Tender Drawings

Specification

Tender Programme

Contract Conditions (where these are not a standard form of contract)

Any correspondence or further details issued during the tender period

Once the contract is awarded a set of contract documents is normally issued, including a copy of the working drawings, any revised specification and the contract conditions. The first task will be to establish if there are any differences between the tender documents and the contract issue of documents and drawings. Any differences are likely to have some implication in respect of material or time and therefore will affect the cost of the work. Any such changes must be recorded and notified to our client on the standard forms, including the implications to cost and programme.

Once we have accepted the contract and advised our client of any changes to the original tender we must ensure that the cost involved in completing the work is controlled. The better the cost control the better the chances of making the necessary level of profit, thus ensuring the job is successful, improving the company's success and securing our own jobs.

Record drawings

Having considered the requirements of the contract we will usually be expected to provide Operation and Maintenance manuals on the completion of the work. Record drawings will need to be included as part of the Operation and Maintenance Manual.

In order that we can carry out our installation we will need a set of working drawings. These will show the proposed route for the services we are to install. During the course of the installation there may be some variation to the routes shown on these drawings. This may be due to construction constraints, other services, access etc. We have a responsibility to provide the client with a set of record drawings showing exactly where the cables and equipment etc. are located.

The usual method of doing this is to provide the supervisor for each area with a spare copy of the working drawing. This is then marked up by the supervisor, on a daily basis, showing the precise location of the installation. On completion of an area this marked up drawing is returned to the draughtsman to allow the finished record drawing to be produced.

This approach has two major advantages
- The detail of the installation is added to the drawing whilst it is still visible. If left until the area is complete, the finishes may already be applied to the early stages, particularly on "fast track" projects. This will then involve the removal of finishes and additional costs to produce our drawings.

- By consulting the "living" record drawing we can easily establish our progress in any area. This is very useful for producing progress reports and monitoring against the programme.

In addition to this it will ensure that the record drawings can be issued on or soon after completion. As these are part of our contract the client is entitled to withhold money until they are released. We may also have to submit our record drawings to the client's consultant for approval before they can be issued. The technique described above allows the record drawings to be approved as the construction progresses, rather than as a "job lot" at the end.

It will also provide a cost-effective use of the draughting facilities as the draughting will also be progressive with a smaller number of staff involved with the project throughout. A much better arrangement than a number of people unfamiliar with the job at the very end.

PROJECT

Now we have completed Chapter 8 of the module we should consider part 5a of the project. Part 5a deals with the documentation associated with instructed changes once we are engaged in a contract.

Answers

These answers are given for guidance and in some instances are not necessarily the only possible solutions.

Where answers have not been given the solution is either an individual one or will be found in the text of this book or the reference book suggested.

Chapter 1

p.4   Try this: see list in Regulation 29 in the Memorandum of Guidance on the Electricity At Work Regulations 1989 (HS(R)25)

Chapter 2

p.9   Try this: Radiation:- coal fire, radiant electric fire, Convection:-transformer tank, convector heater, Conduction:- kettle element, storage heater

p.18   Try this: see p.16 and p.17

p.23   Try this: 3.78 kW

p.27   Try this: $1.589 \text{ W/m}^2 \,^\circ\text{C}$

Chapter 3

p.33   Try this: environmental conditions such as temperature, solar radiation, water, vermin, livestock, corrosive or polluting substances, mechanical impact, flora and mould growth, lightning

p.34   Try this: (1) any area where protection against dust/water spray is required, for example communal shower, flour production area (2) IP45

      Try this: could include: timber, brick, metal sheet, canvas, cast in-situ concrete or steel frame with cladding

p.36   Try this: 108.22 A

p.37   Try this: (1) 12.17 A; (2) 25.97 A

p.41   Try this: see Table 4B1 in BS 7671

      Try this: Increase the correction factor for ambient temperature to be taken into account from 0.71 to 0.87, factor for type of fuse 0.725 needs to be applied, increase in csa of conductor required.

p.42   Try this: should include; No 0.725 fuse factor to be considered, lower ambient temperature factor, minimum I_t reduced, reliability and resetting after fault improved.

p.43   Try this: most onerous situation is where the cable passes through the thermal insulation, correction factor of 0.5 to be applied

      Try this (a) Circuit 1: where single circuit passes through 100 mm thermal insulation. Factors C_f 0.725 and thermal insulation 0.81. (b) Circuit 2: 3 circuits in conduit through thermal insulation. Factor 0.7 and (method 4 for cable selection).

p.45   Try this: (1) 4 mm^2 (2) 2.5 mm^2

p.47   Try this: $R_1 + R_2$ value $0.23\ \Omega$

      Try this: Z_s $0.58\ \Omega$

p.49   Try this: 90 A

p.50   Try this: 1.496 mm^2

Chapter 4

p.63   Try this: (a) should include: acids, corrosive chemicals, sewage, methane gas, rodents, water spray, fungi, purification chemicals and chemical storage areas.

      (b) should include: pumps, motors, dosing units, lighting, power, cleansing pressure sprays, additive units, control valves, timers and extractors.

p.64   Try this: (a) 0.2 seconds; (b) 833 ohms, although values over 100 ohms may be unreliable

p.66   Try this: light switch should be mounted outside the sauna

p.69   Try this: (a) 85 °C; (b) 70 °C; (c) 180 °C

p.70   Try this: (a) 833 ohms, although values over 100 ohms may be unreliable; (b) positioned to avoid risk to livestock from voltage gradients; (c) should include: protection against damage from faeces and urine, attack by rodents, mechanical damage. Requirements for supplementary bonding, lower disconnection times, fire and thermal effects and use of class II equipment; (d) protected by RCD, disconnection time 0.2 sec and suitable for the environment

p.74   Try this: (a) MICC, PTFE, SIR, SBR, IIR; (b) MICC, XLPE, EPR, PTFE, SIR AND Glass

p.76   Try this: (1) zinc and graphite; (2) copper and steel; (3) galvanising, spraying with molten metal, sherardising and electroplating

p.81   Try this: (1) normally open, normally closed; (2) any three of: break glass unit, alarm lever, smoke detector, heat detector, temperature rate of rise detector

p.85   Try this: (1) 12; (2) 6; (3) sirens or horns

p.94   Try this: see p.92

Chapter 5

p.103 Try this: green sleeving omitted on 2 no. cpc, green sleeving too short on 1 no. cpc, exposed conductor outside terminal on 1 no.neutral, exposed conductor outside terminal on flex, cable clamp not used on neutral flex.

p.108 Try this: refer to Section 713 BS 7671

p.111 Try this: similar to:

Instrument function*:	*Insulation resistance*			Acquisition date:		
Make:				Model:		
Serial Number or other assigned identity:	*IR4*					

Location of any designated socket-outlet, or description of any device, used for reference checking purposes:

Resistance box 1

Date:						
Formal calibration certificate	✓			✓		
Check against designated reference circuit/device:		✓				
Cross-check against another test instrument:			✓			
Identity of other test instrument:			*Serial Number IR5*			
Repair:				✓		
Remarks:	*Initial calibration*			*Re-calibrated after repair*		

Full details of the results of the accuracy checks are to be retained for record purposes in support of this summary.
*Separate records are required for each function of a multi-function test instrument

p.118 Try this: see p.116

p.120 Try this: see p.115

p.124 Try this: (1) continuity of cpc; (2) polarity; (3) $R_1 + R_2$ for circuits where Z_s cannot be measured

p.127 Try this: see pp 125 and 126

p.131 Try this: see Figure 5.47

p.136 Try this: limitations of discharge of energy, obstacles, placing out of reach, non-conducting location, earth free local equipotential bonding, presence of undervoltage protection devices where appropriate

p.138 Try this: see p.137

p.139 Try this: (a) see p.138; (b) answer should include: the purpose for which the report is required, e.g. licensing, the requirements of any third party interest, e.g. licensing authority, the extent and limitations to be imposed in agreement with the client, and access and availability of the installation.

SPECIFICATION

Building utilisation

The attached drawings (not to scale) ref D/001, 2, 3 & 4 refer to The Ducan Squash Club. The building contains two squash courts, a fitness room, solarium, changing rooms and a bar with a viewing gallery. A lift is installed for the delivery of goods to the bar area.

Building construction

The building is a two-storey construction with cavity walls, having fair face brickwork for all outside walls. The inner skin is thermal block with appropriate load-bearing sections. The inner walls all have a plaster finish, an appropriate proprietary finish is used within the squash courts. The building has a pitched roof and contained within the roof void is a small plant room.

The roof is constructed of timber trusses with felt and clay tile finish, the roof void is provided with 150 mm thermal insulation between the trusses at ceiling level.

All the floors within the building are cast in situ concrete slab with 50 mm screed. The floor is carpeted throughout with the exception of the squash courts, the fitness room and the solarium which have a gymnasium flooring and the changing rooms and toilets which are ceramic tiles.

The ground floor ceiling is a suspended acoustic tile, 10mm thick, in all areas with a void above of 300 mm. The first floor ceilings are all plasterboard with an artex finish. The ceiling heights are 2.5 m from finished floor level.

All outside windows are to be double glazed units, comprising two sheets of 4 mm glass and a 20 mm air gap.

ELECTRICAL INSTALLATION

This installation is to be carried out in accordance with the current edition of the BS 7671and any statutory regulations, codes of practice and British Standards that may be applicable. As the location is remote there is to be a full electric heating system installed using an economy tariff. All cables used are to be LSF insulated and sheathed where appropriate. Unless otherwise stated all final circuits will be wired with single PVC LSF cables in galvanised steel conduit and trunking.

Electrical supply

The incoming supply is 400/230 V, 50 Hz. three-phase, four wire with measured values of Z_e at 0.05 Ω and I_{pf} 3-phase as 9.2 kA. The system is TN-C-S and the supply intake is protected at the cut out with 200 A BS 88 part 2 fuses.

Distribution

There is a main distribution board at the origin which supplies sub-main distribution cables. These supply distribution boards are located in the office, the bar store and the plant room. Cables are to be copper conductor, LSF insulated and sheathed swa, installed on galvanised steel cable tray. The distribution board for the heating is located at the origin of the installation.

Circuit protection

Type B miniature circuit breakers to BS EN 60898 are to be used for all final circuits.

Heating circuits

These are to be wired in Mineral Insulated Copper Clad Cable with an Orange LSF sheath.

Bar circuits

All cabling and accessories associated with the bar, including the dispensing equipment, will be supplied by a specialist installer. A total load of 80 A has been requested.

Fire alarm

To be determined as part of the project, although the installation will be carried out using MICC with a Red LSF oversheath throughout.

Emergency lighting

Self-contained emergency luminaires are to used.

Telephone and security systems

To be fitted by specialist installers. A dedicated circuit rated at 20 A is to made available at the origin.

ASSIGNMENTS

You are in charge of the installation of electrical wiring and equipment at the Squash Club site. You are also responsible for the design of the installation for the scheme. You have the authority to liaise with the Architects, the clients agents and the clerk of works and suggest any changes to the scheme that you think may be necessary. You will need to use the drawings provided to complete these assignments.

1   (a) List the statutory and non statutory regulations which will be applicable to the design and installation of the electrical installation for the squash club site.
     (b) Detail the specific statutory regulations relating to the requirements for working on electrical equipment that has been made dead. List the items that must be achieved in order to comply with the requirements.

2   (a) Describe one method of reducing heat loss from the building due to the opening and closing of the doors.
     (b) Suggest a suitable type of frost protection heater for the plant room using manufacturers' catalogue details.

3   Calculate the minimum size of the sub main cable to the distribution board, mounted at a height of 1.8m, in the roof plant room for compliance with BS 7671 if the circuit length is 22 m. The cable is to be a copper conductor, LSF/SWA/LSF cable and the load to be assumed for the plant room is 3 phase at 60 A per phase. The ambient temperature for the cable run is 40 °C and the cable is installed spaced from other cables. It is intended to use the cable armour as the CPC.

4   (a) Draw a distribution diagram for the installation
     (b) List, in the correct sequence, all the tests that will need to be carried out at the main intake position.
     (c) Describe, in detail, the tests that would be carried out to establish Z_{DB} for the first floor distribution board.

5   (a) At the beginning of the contract we are given an instruction that the night storage heaters, shown on the drawings, are to be changed to under floor heating throughout the building. Detail the documentation that you would expect to receive and issue in order to carry this instruction out.
     (b) Produce a detailed and priced list of the material content for the omission of the storage heaters. (The heaters need only be included as the number of heaters and an as per schedule comment.)

6   (a) Produce a simple bar chart to show the sequence of activities required to install the lighting in the squash courts. You are to be allowed clear access for a period of eight working days during which the installation is to be completed. This will include the erection and disassembly of all plant and equipment as well as the electrical installation. On the bar chart identify the milestone times for the delivery of plant and materials.
     (b) Based on the bar chart for the squash lighting, produce a simple labour histogram for the installation period.

7   (a) Show the location of the fire alarm detectors, break glass units and sounders and control panels for the building. State the type of device installed in each case.
     (b) List any other British Standards or codes of practice which may apply to the fire alarm installation.

ELECTRICAL SYMBOL LEGEND FOR C.T. SQUASH CLUB

	1800 mm LUMINAIRE
	1500 mm LUMINAIRE
	1200 mm LUMINAIRE
	RECESSED CIRCULAR LUMINAIRE
	600 × 600 mm RECESSED MODULAR LUMINAIRE
	WALL LIGHT
EM	EMERGENCY LUMINAIRE
A	50 W 12 VOLT DOWNLIGHTER WITH BRASS FINISH
B	2 × 18 W COMPACT FLUORESCENT LUMINAIRE
C	2 × 9 W COMPACT FLUORESCENT UP/DOWNLIGHTERS
D	2 × 18 W COMPACT FLUORESCENT LUMINAIRE c/w 3 hr EMERGENCY BATTERY PACK
E	28 W 2D FLUORESCENT EXTERNAL LUMINAIRE c/w 3 hr EMERGENCY BATTERY PACK AND PHOTOCELL
F	28 W 2D FLUORESCENT BULKHEAD TYPE LUMINAIRES c/w 3 hr EMERGENCY BATTERY PACK
F2	SINGLE 1200 mm 36 W SLIMLINE FLUORESCENT LUMINAIRE
F3	1200 mm 36 W SINGLE SEALED IP54 FLUORESCENT LUMINAIRE
F4	1200 mm 36 W SINGLE SURFACE MOUNTED PRISMATIC LUMINAIRE
F5	1200 MM 36 W LUMINAIRES WITH METAL TROUGH REFLECTOR
F6	1800 mm 70 W SINGLE SURFACE MOUNTED PRISMATIC DIFFUSER LUMINAIRE
F16	1800 mm 70 W SINGLE SURFACE MOUNTED REFLECTOR LUMINAIRE WITH WIRE GUARD
G	50 W 12 VOLT DOWNLIGHTER WITH WHITE FINISH
	LUMINAIRES WITH THE LETTER "E" ADJACENT TO BE FITTED WITH 3hr SELF CONTAINED EMERGENCY PACK
	PULLCORD SWITCH
	1 WAY SWITCH
	2 WAY SWITCH
	INTERMEDIATE SWITCH
T	TOKEN OPERATED TIMER SWITCH
	DISTRIBUTION BOARD
M	METER
	TP & N ISOLATOR
FA	FIRE ALARM PANEL
	FLEX CABLE CONNECTION POINT
	SWITCHED 13 A FUSED CONNECTION UNIT
	DOUBLE 13 A SOCKET OUTLET
	SINGLE 13 A SOCKET OUTLET
A	110 V SINGLE SOCKET OUTLET
NSH	NIGHT STORAGE HEATER UNIT

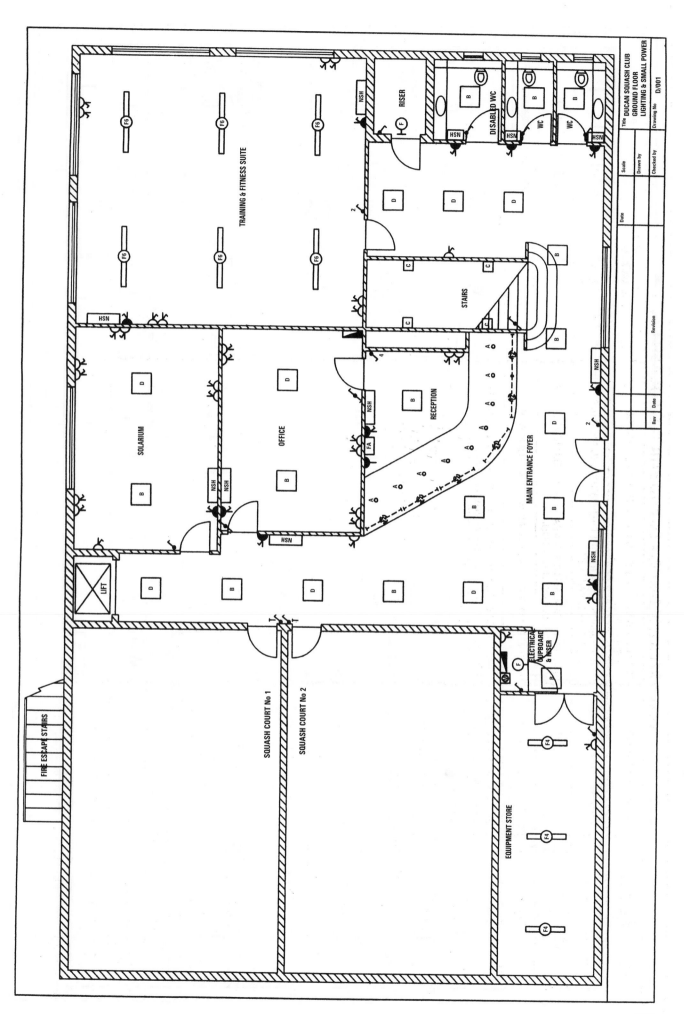

177

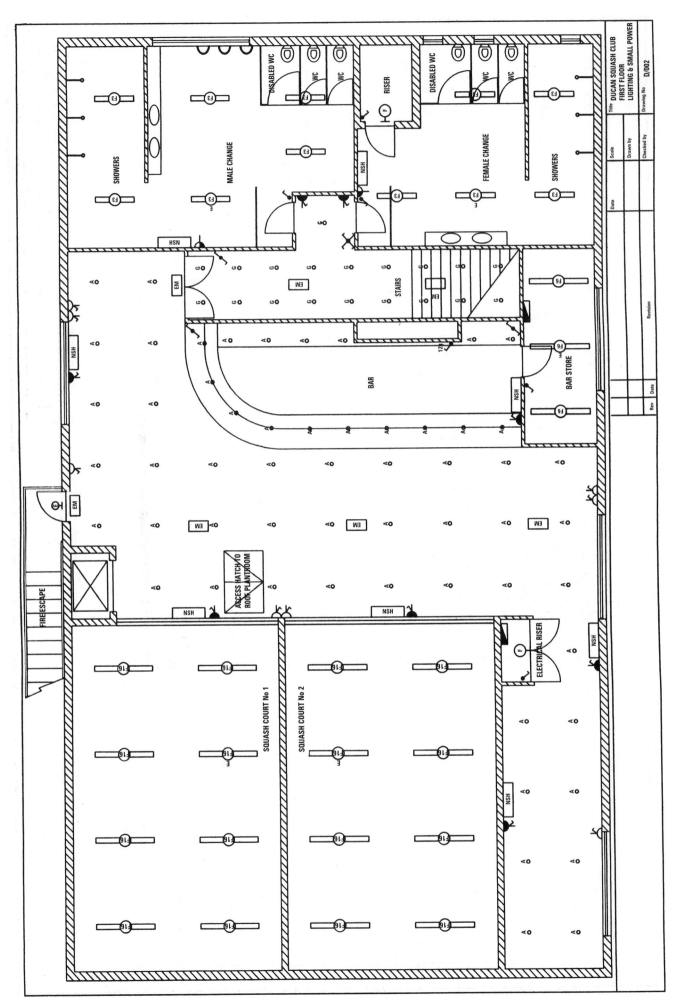

178

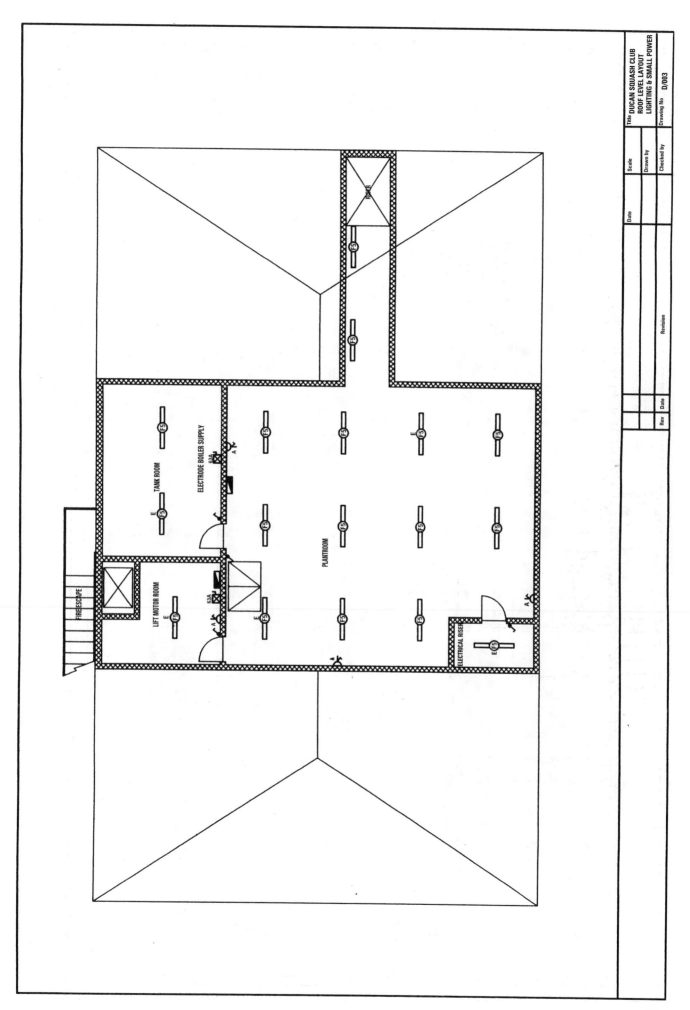

FIRESCAPE

LIFT MOTOR ROOM

TANK ROOM

ELECTRODE BOILER SUPPLY

PLANTROOM

ELECTRICAL RISER

Title **DUCAN SQUASH CLUB**
ROOF LEVEL LAYOUT
LIGHTING & SMALL POWER

Drawing No **D/003**

Scale

Drawn by

Checked by

Date

Revision

Rev | Date

179

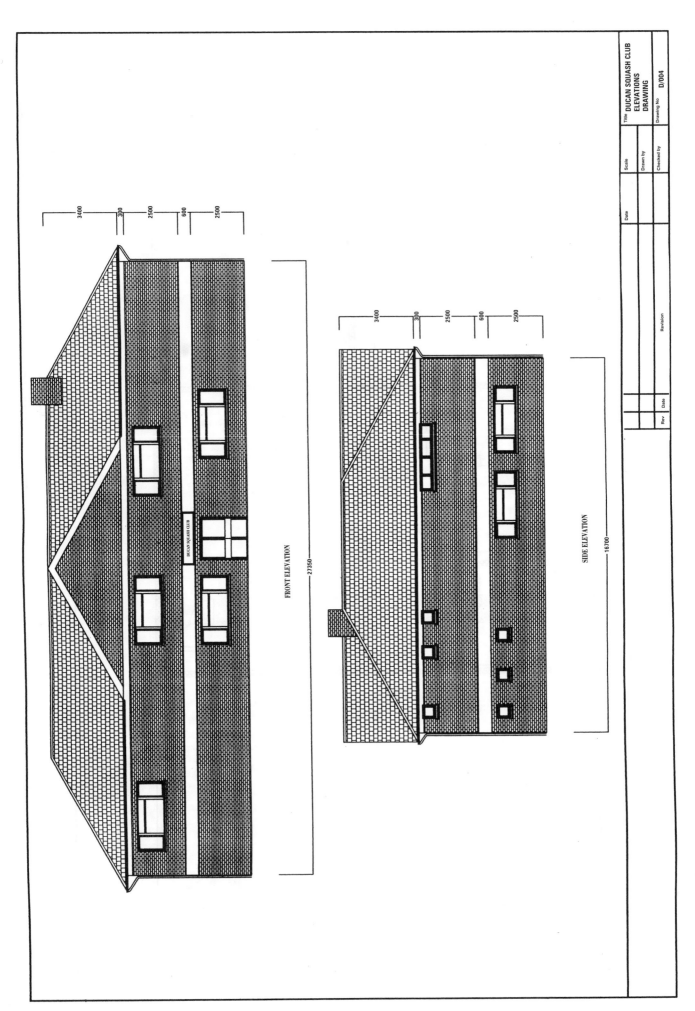

FRONT ELEVATION

27350

3400 300 2500 600 2500

SIDE ELEVATION

16700

3400 300 2500 600 2500

DUCAN SQUASH CLUB

Title DUCAN SQUASH CLUB
ELEVATIONS
DRAWING

Drawing No   D/004

Scale

Drawn by

Checked by

Date

Revision

Rev   Date